Circadian Clock and Aging

Syed Ibrahim Rizvi • Gaurav Majumdar

Circadian Clock and Aging

Springer

Syed Ibrahim Rizvi
Department of Biochemistry
University of Allahabad
Allahabad, Uttar Pradesh, India

Gaurav Majumdar
Department of Zoology
University of Allahabad
Allahabad, Uttar Pradesh, India

ISBN 978-981-97-6694-9 ISBN 978-981-97-6695-6 (eBook)
https://doi.org/10.1007/978-981-97-6695-6

This Springer imprint is published by the registered company Springer Nature Singapore Pte Ltd.
The registered company address is: 152 Beach Road, #21-01/04 Gateway East, Singapore 189721, Singapore

Preface

Our scientific knowledge suggests that Earth is likely the only celestial body in the vast Universe where life exists. The life which we as humans perceive is based on a set of reactions that happened spontaneously between different basic biomolecules and the same set of reactions then created a group of complex structures, each of which was capable of creating a similar structure or was capable of reproduction. The size, shape, and complexity of the resulting structure may vary, but the common denominator is that anything which can be considered living should be able to display many characteristics of which the most important is the ability to reproduce.

Life originated on this blue planet some 3.8 billion years back. Random mutations and adaptation driven by natural selection have helped the diversification of life forms from bacteria to mammals. It may look daunting to explain this journey of life solely on the basis of natural selection theory of Charles Darwin, but all the justification against this theory is based on conjectures. It is beyond the capacity of human mind to fathom the span of time that has passed since life evolved from a single bacterial cell to its present-day diversification.

In the backdrop of the origin of life on this planet and its journey to the present moment, all life forms at all times have been exposed to earth's atmosphere and earth's rotation on its axis, thereby creating distinct phases of light and dark periods. The cyclic change of light and dark periods has been inherent to the survival of all living organisms, including humans. It is known that living organisms undergo cyclic variations in many of their basic functions including physiology, biochemistry, and behaviour. These sets of cyclic variations are clubbed as circadian rhythms.

Since living organisms have evolved under the influence of day/night routine, any change or alteration in this daily rhythmic pattern becomes a cause of concern. The living system experiences "stress" when subjected to a change in circadian rhythm, although sometimes the "stress" may not be visible in a short time scale. Research in the area of circadian oscillations and its effect on human behaviour, physiology, biochemistry, and psychology have provided many startling results. It is now acknowledged that humans are badly affected by changes in circadian rhythm, and these changes may hold a clue to various diseases. The most startling fact that has emerged is that circadian rhythm may also affect human aging.

The relevance of the research on circadian rhythm is becoming more and more important since human technological advancement has created an environment where the distinction between day and night has virtually vanished. The concept of

"Artificial Light at Night" or ALAN is being intensely investigated for its effect on human life. The change has been so fast in human history that natural selection has had no opportunity to adapt to such changes in circadian rhythm. Evolving data shows that human technological advancement and changes in circadian rhythm have also affected animals.

The last two decades have seen exponential increase in research on the mechanisms of aging. Biogerontologists have now more or less agreed on the hallmarks of aging. With a clearer understanding of the factors which play a role during aging, the next frontier is now to work on anti-aging interventions. The relationship between aging and circadian clock is still not fully understood. We present our views on the involvement of circadian clock in the aging process.

This book is divided into eight chapters which start with a detailed introduction about circadian rhythms and then move on to circadian oscillation and then the concept of biological clock. Beyond these introductory chapters, which provide the concepts, we move on to models to study circadian system. We explain in detail how circadian rhythm affects our brain and associated pathological states. The following chapters deal with how circadian rhythm affects aging and whether a tweaking of the circadian system in the present-day scenario may be a possible anti-aging intervention strategy.

It has been our continued effort to provide a concise present-day scientific knowledge about circadian clock and how it is affecting human health with respect to human aging.

Allahabad, India Syed Ibrahim Rizvi
Allahabad, India Gaurav Majumdar

Contents

About the Authors

Syed Ibrahim Rizvi is a Professor of Biochemistry and Dean of Research and Development at the University of Allahabad, India. He is a Visiting Professor at Ohio State University (USA), University of Rome, University of Milan, University of Athens, University of Nice, University of Szeged, University of Pisa, Karolinska Institute, Slovak Academy of Sciences, Istanbul University, and Bangladesh Institute of Health Sciences. He has also been an invited speaker at major conferences in the USA, Russia, Hungary, Italy, Saudi Arabia, Kuwait, Egypt, Bangladesh, and Thailand. Prof. Rizvi has done seminal research work in chronobiology, mechanism(s) of aging, antiaging interventions, and lifespan. He has published extensively in high impact factor journals. To date, Prof. Rizvi has published more than 200 research papers. He serves on the editorial board of the prestigious Springer journal Biogerontology. Prof. Rizvi has been a recipient of grants from the International Cell Research Organization, Federation of European Biochemical Societies, European Molecular Biology Organization, Organization for the Prohibition of Chemical Weapons, International Foundation for Science, Department of Biotechnology (Government of India), and Department of Science and Technology (Government of India). Prof. Rizvi was conferred the Academic Excellence Award by the University of Allahabad in 2016. Prof. Rizvi has also written many popular science articles which have been published in national newspapers. He is a member of Indian Gerontological Society, Society of Biological Chemists, National Academy of Sciences India, and Asian Network of Research on Antidiabetic Plants.

Gaurav Majumdar is working as Assistant Professor of Zoology at University of Allahabad, Allahabad. He did his Ph.D. in chronobiology from the University of Delhi. His research has focused on the role of circadian and seasonal factors in regulating avian migratory physiology. After his Ph.D., Dr. Majumdar acquired two prestigious European post-doctoral research fellowships (BELSPO and FWO) to continue his post-doctoral research at the Bio-Imaging Laboratory, University of Antwerp, Belgium, where he worked on the role of light, opsin molecules, and circadian clock in mediating neurodegenerative diseases with an emphasis on Alzheimer's disease. He then moved to the University of Glasgow to work on the molecular mechanisms involved in seasonal clock regulation of animal physiology. He has published extensively on circadian clock and its interphase with seasonal physiology of vertebrates. Dr. Gaurav is an executive member of Indian Society for Chronobiology (InSC) and Association of Avian Biologists in India (AABI). Recently, he was selected as Young Associate (2024) by Indian National Science Academy (INSA). He is the son of Snigdha and Sibdas, husband of Garima and father of shinoy.

Introduction to Circadian Rhythms 1

1.1 Introduction

The field of chronobiology, which focuses on the circadian rhythm, which is a rhythmic cycle of approximately 24 h, has undergone a significant transformation in our comprehension and practical implementation throughout the last five decades. We have transitioned from disputing the presence of an internal "clock" to identifying a significant genetic mutation in the human version of a gene found in fruit flies that controls the regularity of behaviour and pathology. Before, we only debated whether or not an endogenous "clock" actually existed. Obviously, such a burst in the body of known information does not occur de novo or by random coincidence. Who were the people who "equipped the brains of modern researchers for discovery" by making the observations, experiments, and realizations that came before them?

The following is a brief account that attempts to highlight the scientific path that preceded and launched modern mechanistic research in circadian biology. This line of inquiry, according to our own personal estimation, dates back to the 1970s; the astounding progress that has been made since then is clearly described in the chapters of this volume. There are already important accounts of the history of chronobiology as well as the memoirs and comments of individuals who made it, such as Erwin Bünning, Jürgen Aschoff and Colin Pittendrigh, which can be found in the existing body of literature.

1.2 Evolution of Circadian Rhythmicity

Greeks had two words for time. Chronos (ovo) is linear time from the past to the future. Kairos, or timing, is the proper time to act. "Observe due measure, for right timing is in all things the most vital factor", reads Hesiod's poem Works and Days, c. 700 BCE. So, thousands of years before present compilations of the best times of

S. I. Rizvi, G. Majumdar, *Circadian Clock and Aging*,
https://doi.org/10.1007/978-981-97-6695-6_1

day for various behavioural and physicochemical functions, biological and geophysical time were differentiated and optimal phases under entrained conditions were understood.

Aristotle (384–322 BCE) emphasised the significance of rising early in the morning, as it promotes physical well-being, prosperity, and intellectual acumen. This ancient affirmation of the advantages of being an early riser may have sprung from much earlier practices of worshipping the sun and the desire to welcome the first light of day. The quotation “Early to bed and early to rise, makes a man healthy, wealthy, and clever” by Benjamin Franklin (1706–1790) has been handed down for centuries in English-speaking countries, along with the works of Samuel Johnson (1709–1784; *I have, all my life long, been laying till noon; yet I warn all young men, and tell them with great earnestness, that nobody who does not rise early will ever accomplish any good)*. Theophrastus of Eresos, who is considered the pioneer of botany, documented the observations made by Androsthenes of Thasos during his trip from the Indus River to the Persian Gulf following Alexander the Great’s Indian campaign. These discoveries were recorded in Theophrastus’ work, Historia Plantarum, which was written between 326 and 324 BCE. Androsthenes provided a description of the tamarind tree in Tylos (now Bahrain) as a tree that has a large number of leaves. He observed that the tree shuts its leaves at night and opens them at daybreak. By noon, the leaves are fully unfolded. In the evening, the tree gradually closes its leaves again and remains closed throughout the night. The locals believe that the tree goes to sleep during this time. Throughout history, researchers have observed the phenomenon of nyctinasty, which refers to the daily motions of leaves. However, it wasn’t until the seventeenth century that experiments specifically studying this behaviour were published. Prior to presenting his celebrated experiment to the Royal Academy of Sciences in Paris in 1729, Jean-Jacques d’Ortous de Mairan (1678–1771) had already produced dissertations on physics, mechanics, and optics. He gained high esteem, was chosen for leadership positions in the Academy, and was admitted into prestigious European learned groups. At that time, he was regarded as one of the most exceptional scientists of the century, and in 1935, a crater on the moon was named after him, called Mairan. De Mairan positioned “sensitive plants” in a light-deprived location, shielded from sunlight and air, as documented in a concise “botanical observation” (Emery et al. 2023). The plants exhibited diurnal leaf movements, opening and closing their leaves in a cyclic pattern. Therefore, although de Mairan should be acknowledged for moving a plant away from sunlight, his contribution to chronobiology is primarily based on a publication that lacks details on the absence of light and the specific timing of the observations. It was not stated that the plant unfurled its leaves in the absence of light by relying on an internal timekeeping mechanism. The plant “perceived the sun without visual perception”, hence de Mairan inferred an external origin.

Henri-Louis Duhamel du Monceau (1700–1782) in France, John Hill (1716–1775) in England, and Johann Gottfried Zinn (1727–1759) in G€ottingen (then in the Kingdom Hannover, now Germany) independently studied de Mairan’s paper in the following years. De Mairan’s primary discovery was confirmed, and further research demonstrated that the experience of daylight is not influenced by light leakage or temperature (Emery et al. 2023). Augustin Pyramus de Candolle, a Swiss botanist born in 1778 and died in 1841, achieved significant breakthroughs in his field throughout the early 1800s. He was the initial member of a lineage of four de Candolle botanists, and the journal Candollea, published by the Conservatory and Botanical Garden of the City of Geneva, is named in his honour. De Candolle conducted a study on the leaf motions of Mimosa pudica in constant light and found that the duration of "sleep and wakefulness" was less than 24 h and differed among plants ("in certain pots, the acceleration reached one and a half to 2 h each day") (Schwartz and Daan 2017). De Candolle deduced that the sleep-wake movements of the plant are connected to an intrinsic tendency for periodic motion based on the deviation from a 24-h cycle. The concept of an internal origin for the periodicity was first proposed in 1832. Several prominent botanists, such as Julius Sachs (1832–1897) (Schwartz and Daan 2017), agreed with De Candolle's stance. However, Wilhelm Pfeffer (1845–1920), a respected German plant physiologist with a wide range of interests, including thermonasty and photonasty, did not find De Candolle's arguments convincing. "The experiments presented in this study provide undeniable evidence that the daily periodic movements cannot be regarded as historically established characteristics, as they gradually cease under continuous illumination", the author stated, referring to these movements as "post-oscillations" of a system returning to a stable state after being disrupted. Charles Darwin (1809–1882) held the belief that everyday movements were inherited from inside the organism itself. Merely 5 years later, he collaborated with his botanist son Francis to co-author, *The Power of Movement in Plants*, wherein they proposed a specific recommendation for the benefits of selection.

The case for an endogenous origin of daily rhythmicity, or at least a component of it, kept growing as scientific evidence overtook philosophical arguments as the twentieth century went on. In retrospect, Anthonia Kleinhoonte's (1887–1961) studies on Canavalia ensiformis, the plant that produces jack beans, were exceptionally insightful. The seedlings were initially kept in an 8:8 h light-dark cycle. Still, she showed in the 1920s that leaf movements exhibited a circa 24 h rhythm in constant darkness regardless and that brief light pulses, as short as one minute in duration, could reset the rhythm phase depending on the phase of light administration. Plants could be made to show various phases in the absence of light, refuting the idea that these rhythms were a direct result of an unidentified foreign component. Although he began his career as a legitimate botanist at the University of Jena in Germany, Erwin Bünning (1906–1990) could be regarded as the first genuine circadian biologist. By giving circa 24-h rhythms a name (endodiurnal; Franz Halberg (1919–2013) popularised the term circadian in 1959 (Halberg 1960), he contributed to its legitimacy. He also authored what is perhaps the first approachable "textbook" on the subject (Die Physiologische Uhr 1958). The fact that his research included both the plant and animal kingdoms and was not limited to a single organism is possibly the most significant factor. He demonstrated in the 1930s that endodiurnal rhythmicity in Phaseolus was clearly and significantly longer than 24 h, that cultivating the plants in abnormal light-dark or constant conditions had no impact on the periodicity of succeeding generations, and that researching hybrids of plants with different phenological traits implied a genetic basis. Subsequently, he conducted research on the 24-hour cycle of adult fruit flies (Drosophila) that emerge from pupae (a process called eclosion). His findings demonstrated that this rhythm persists in the progeny of flies that have been exposed to continuous dim light for multiple generations. It is worth noting that similar work was also carried out by Hans Kalmus (1906–1988) in Prague.

As the middle of the twentieth century drew near, it became evident that 24 h biological rhythms were a widespread phenomenon covering a wide range of functions in various animals (albeit endogenous generation had not yet been widely accepted). This also applies to vertebrates. Curt Richter, an American psychobiologist (1894–1989), collected data on rat locomotor activity during the late 1920s. This data included information from rats that had access to "revolving drums" (running wheels) which were created a few years prior by Colin Stewart at Clark University in Worcester, Massachusetts. Body temperature has become the preferred method of testing for larger species, including humans. It is also a natural choice for a new group of researchers who have acquired skills in physiology and medicine. Sutherland Simpson (1863–1926), a Scottish physician, courageously documented the axillary temperature cycles of rhesus monkeys (Macaca mulatta). He subsequently moved his physiological laboratory to Cornell University in New York. He performed this task in light-dark, reversed light-dark, and constant dark conditions. American physiologist Nathaniel Kleitman (1895–1999), who is regarded as the father of contemporary sleep research, became intrigued by the regulation of sleep and wakefulness timing. He and his aide Bruce Richardson spent 32 days alone in Mammoth Cave, Kentucky, in 1938, trying to become used to a 28-h day (19 h awake with the lights on and 9 h in bed with the lights off) at a steady 54 °F (12.2 °C). Their body temperature readings, which were eventually published, suggest that the two guys expressed themselves at different times. Coincidentally, he and his Doctoral student Eugene Aserinsky (1921–1998) disproved the idea that a calm, resting brain occurs during sleep decades later by discovering REM sleep [ref]. The eldest son of the well-known German cardiac pathologist who gave his name to the Aschoff-Tawara node (now known as the atrioventricular node) and Aschoff bodies, Jürgen Aschoff (1913–1988), was born (myocardial nodules following rheumatic fever). The younger Aschoff, who also studied medicine, looked into human thermoregulation physiology in 1939. Through self-experimentation, he found that in constant conditions without food intake, sleep, or movement, there was an unrecorded daily cycle of heat loss. He developed a fascination for the topic and, while working with rats, birds, and people, as well as broadening his knowledge to include ethology and applied mathematics, established a thorough oeuvre on the characteristics and entrainment of physiological rhythms. Later, in 1961, he had the opportunity to conduct research on nine people who spent between two and four weeks living in isolation without access to clocks or other timing devices in a civil air-raid shelter located underneath the university's surgery clinic in the heart of Munich. Aschoff showed that all individuals had ongoing rhythms with periods longer than 24 h (for details on the topic, see (Schwartz and Daan 2017). Aschoff constructed an underground bunker near his institute in Erling-Andechs (less than

50 km from Munich) expressly to serve as a temporal isolation facility where two people may be investigated concurrently after this initial discovery of human circadian rhythms. This sparked a series of groundbreaking studies conducted over 25 years on close to 450 participants that would become the gold standard for studies of human circadian rhythmicity. Aschoff was able to bring the field's new insights to the attention of the world's biomedical community by acting as its enthusiastic, animated, and dramatic advocate.

1.3 The History of Finding Clock in Rhythms

Figure 1.1 illustrates key events in circadian biology in a concise manner. Some individuals hypothesised that creatures might utilise their internal "clocks", which generated visible 24-h oscillations, to accurately measure time well in advance. Julien-Joseph Virey (1775–1846) worked as the chief chemist at the Val-de-Grace hospital in Paris before obtaining an MD degree in 1814. He documented the presence of a particular kind of living clock, which he referred to as "une sorte d'horloge vivante"—entrainée (entrained), after monitoring the cyclical patterns of clinical phenomena at the hospital, that are influenced by the movements of the sun and earth. In 1939, the American zoologist Maynard Johnson published activity records (cage movements) of deer mice (Peromyscus leucopus) in continuous light. He concluded that these animals possess a remarkably robust and long-lasting physiological clock that winds and regulates itself. However, the specific mechanism of this clock is yet to be determined. In 1936, a 30-year-old Bünning suggested a method for incorporating a 24-h clock into a device to measure daylight's fluctuating length (Schwartz and Daan 2017). Johnson left the Biological Laboratory of Harvard University and went to the Lake Mattamuskeet Wildlife Refuge in North Carolina, where he made his last contribution to the field of circadian biology.

The concept of conceptualising rhythms as clocks started taking shape in the 1950s. To accurately estimate external time, an internal clock needs to establish a stable phase relationship with the environment, encompassing both the time of day and the passage of time. The American biologists J. Woodland Hastings and E. Beatrice Sweeney measured the circadian bioluminescence rhythm of a photosynthetic dinoflagellate called Gonyaulax polyedra (now known as Lingulodinium polyedrum). Additionally, Patricia De recorded the circadian wheel-running rhythm of the flying squirrel, Glaucomys volans. The "phaseresponse" curves obtained by Schwartz and Daan (2017) showed that a combination of an internal oscillation and a light-responsive rhythm could lead to exact and accurate synchronisation with the natural light-dark cycle. Colin S. Pittendrigh (1918–1996) obtained his undergraduate degree in botany from the University of Durham in England. He was dispatched to Trinidad during World War II to engage in malaria control efforts. It was during this time that he initially witnessed the daily activity patterns of mosquitoes. He obtained his PhD in genetics and evolutionary biology from Columbia University in New York in 1948, studying under the guidance of the distinguished geneticist Theodosius Dobzhansky. Dobzhansky is famous for his statement "nothing in

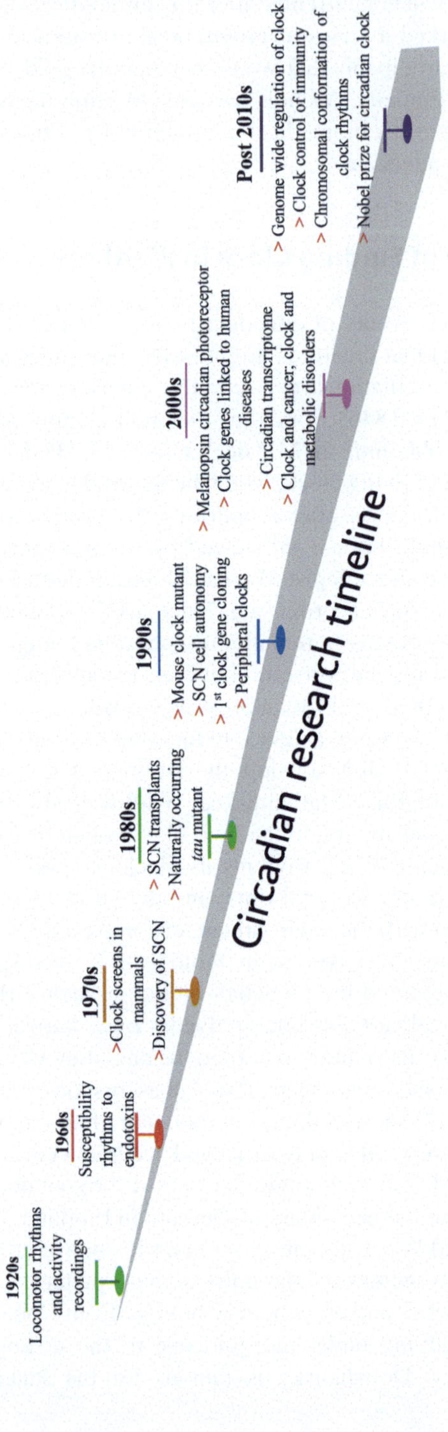

Fig. 1.1 Important timeline events in chronobiology research

biology makes sense except in the light of evolution". Pittendrigh remained in the nation and substantially influenced circadian biology and the development of the "clock" metaphor, even after his official retirement in 1984, when he relocated to residences in Arizona and Montana. Pittendrigh asserted that a reliable clock must exhibit temperature independence, meaning that the circadian period remains constant regardless of temperature. This principle should also apply to the spontaneous frequency of any sensory organ if it is to be used for functions other than temperature sensing. In contrast, Kalmus argued that the period of the Drosophila eclosion rhythm is influenced by temperature. In a meticulous and thorough set of experiments, the researcher examined the rhythm and the impact of light and lack of oxygen on the timing of eclosion in Drosophila pseudoobscura. The findings revealed that the rhythm is not significantly influenced by the surrounding temperature, contradicting Kalmus' mistaken conclusion, which was based on measuring only the initial cycle following a change in temperature. The circadian clock operates independently from the activities it controls and serves the purpose of synchronising an organism's timing with the environment. Additionally, it plays a role in organising the temporal patterns of many subsystems inside the body. Pittendrigh collaborated with others to provide the foundation for these and subsequent investigations on flies and nocturnal rodents. At the age of 45, he became the inaugural circadian biologist to be chosen as a member of the US National Academy of Sciences. He possessed great importance and had a keen insight. Teaching, particularly to students, played a vital role for "Pitt" (as he was referred to by his colleagues). He was an enthusiastic and captivating speaker, often pacing the floor and usually perspiring heavily by the conclusion of his lectures. Pittendrigh acknowledged that the eminent German biologist and ornithologist Gustav Kramer (1910–1959) had impacted his conception of an innate, beneficial circadian clock. Kramer elucidated the utilisation of solar cues in the seasonal migration of European starlings *(Sturnus vulgaris*), enabling them to maintain their intended trajectory. Kramer conducted ingenious experiments utilising circular cages and mirrors to manipulate the perceived position of the sun in the sky. Through these experiments, he identified the presence of a time-compensated sun compass, which involves using the sun's direction as a navigational guide while considering its movement over the course of the day (Campbell et al. 1999). Subsequent research altered the birds' internal clocks and then subjected them to sunlight to observe consistent deviations in their flight trajectories, which were interpreted as evidence of a circadian mechanism (Campbell et al. 1999). Finally, a laboratory setup located a few hundred metres from the South Pole employed turntables that rotated once every 24 h in the opposite direction of the earth's rotation, effectively nullifying it. This allowed researchers to confirm the presence of circadian rhythmicity in hamsters (Mesocricetus auratus), fruit flies (*Drosophila pseudoobscura*), bean plants (*Phaseolus vulgaris*), and fungi (*Neurospora crassa*) (Schwartz and Daan 2017).

1.4 Features of the Rhythms

Like the prerequisites for a majority of scientific disciplines, certain words must be understood to fully comprehend the properties of biological rhythms. The study of the temporal aspects of biological events is known as chronobiology in the case of biological rhythms. This term was first used in the middle of the 1960s (Cambrosio and Keating 1983; Halberg 1960). A lexicon of some of the more typical terminology used in chronobiology to define rhythms is shown in Table 1.1.

The initial aim of a clock biologist/enthusiast is to define or explain how basic terminology like rhythm, period, phase, and amplitude relate to all rhythms, whether they are strictly mathematical, physically based, or biological. These terms offer a means to classify biological cycles (e.g., if a cycle is finished in a few seconds, 1 h, 1 day or 1 year) and are usually stated in numerical units like time or angular degrees. A change that is repeated with the same pattern, likelihood, and period is referred to as a rhythm. The observed oscillation can be categorized as a biological rhythm if the variable that fluctuates is biological and endogenous (driven from inside as opposed to merely being a response to the environment). Three characteristics are frequently used to explain rhythms in biological variables, such as the movements of leaves on a plant or changes in body temperature in a human. The ***period, amplitude, and phase*** are these distinguishing characteristics (Fig. 1.2). Each can be stated in numerical units, giving the rhythm a way to be measured.

Time and frequency: The duration of a cycle is indicated by the period (τ). It could be anything from a few milliseconds to several centuries. The reciprocal of the period ($1/f$), frequency (f) is the number of cycles in a unit of time. A behaviour

Table 1.1 Glossary of important clock-related terms

Term	Description
Biological rhythms	Changes in a biological variable that recur with a similar pattern and interval
Circadian time	Time that spans the circadian period in relation to the light-dark regime
Cycle	A recurring pattern
Entrainment	Coupling of the period and phase of a rhythm with another cycle
Zeitgeber	Environmental signal or cycle that entrain a bio rhythm
Free running	Desynchronization of the period of a rhythm from the period of a known *zeitgeber* under constant condition
Damping	Decrease in amplitude of a rhythm over time
Masking	Superficial change of rhythm characteristics by external environmental conditions
Acrophase	Peak of a mathematical curve fit to data
Amplitude	Distance from rhythmic mean to the peak or to the trough of a mathematical model
Frequency	The number of cycles per unit of time
Period	Duration of one complete rhythmic cycle
Phase	Indicates the location in time for a value within a rhythm
Phase-shift	A change in the timing of the phase, a rhythm to occur earlier or later

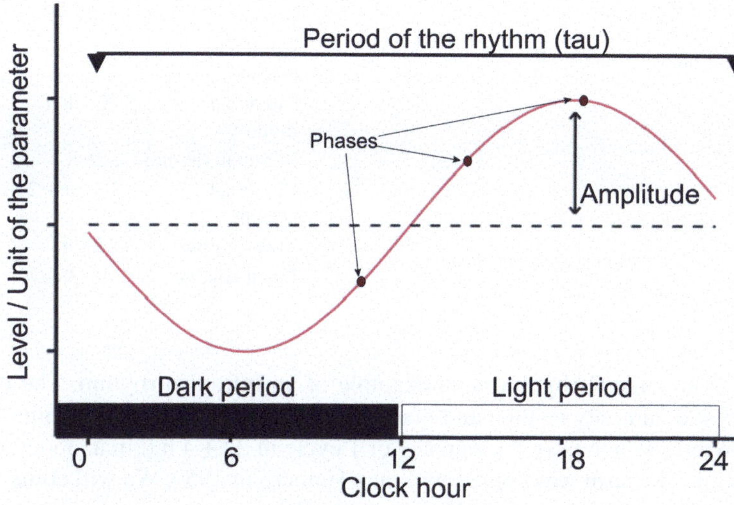

Fig. 1.2 Representation of basic properties of circadian rhythm

that oscillates four times an hour, for instance, shows a frequency of four cycles per hour, which is equivalent to a period of 15 min.

Amplitude: The amplitude of a variable used to define a rhythm is its magnitude from the baseline (often the middle value or mean) to the peak or trough of the mathematical model (for example, a cosine). It is necessary to identify an amplitude to discern a rhythm. Because it depends on the strength of the biological signal and the sensitivity of the monitoring equipment, the value of the amplitude can occasionally be challenging to interpret. A biological variable's amplitude is expressed in conventional quantities as mg, cm, beats/min, or mmHg. When comparing rhythms for other variables, these units might need to be standardised or altered in some way, such as by expressing each value as a percentage of the general mean.

Phase: The phase is yet another aspect of a rhythm. In general, a phase can be thought of as any instantaneous repeated cycle condition (for instance, consider the Moon's phases). In addition to the trough and any intermediate points between the peak and trough, the peak is a distinct point in a cycle that can be referred to as a phase. The acrophase is a term used to refer to the time interval between the peak phase of a mathematical curve and a specific reference point. This reference point can be midnight (00:00 h), the time of midsleep, the time of light onset or awakening, or any other chosen point. The acrophase is calculated using a cosine curve to measure rhythm characteristics. The bathyphase refers to the dip of the cosine curve that has been adjusted to fit the data.

The period domains: Biological rhythms are usually distinguished or grouped in reference to the length of their period. The rhythms according to the periods they display are shown in Table 1.2.

Out of the rhythms detailed in the table above, the three broad domains that are frequently used to classify rhythms are ***ultradian, circadian, and infradian***. For

Table 1.2 Time domains used to categorize different periods of rhythms

Domain	Period
Ultradian rhythm	<20 h
Circadian rhythm	20–28 h
Infradian rhythm	>28 h
Circanual rhythm	1 year ± 2 months
Tidal rhythm	24.8 h
Lunar rhythm	29.53 days

example, a circannual rhythm is an example of an infradian rhythm. The current book's most commonly studied and main focuses are the circadian rhythms. These are the rhythms which have a frequency of 1 cycle in 24 ± 4 h (circa: about; dies: a day or 24 h). The term was coined by Franz Halberg in 1959. We will come across numerous examples of circadian rhythms in this book.

1.5 Ultradian Clocks: Small Steps Toward Circadian Oscillators?

If a biological rhythm has a shorter time than 24 h, it is referred to as ultradian. In addition to behavioural activities, which are frequently connected to food habits, ultradian rhythms have been detected in physiological processes such as cellular processes, respiration, circulation, hormone release, and sleep stages. Ultradian rhythms differ not only in their period duration (from hours to milliseconds) but also in their mechanics and purposes. Generally speaking, the roles of ultradian rhythms have been explained in terms of internal coordination and energy optimization. Ultradian rhythms fall into at least four distinct types due to its broad categorization. Ultradian rhythms are continuous and continue throughout the whole 24-hour episode in their most basic form. Yet, most animals exhibit polyphasic (and hence ultradian) activity patterns confined to particular circadian times of the day. In addition, activity bursts can occur at highly exact intervals within the circadian rhythm, such as bimodal or trimodal activity at the beginning, middle, and end of the activity phase. These regular intervals between episodes are referred to as ultradian rhythms (Poon et al. 1997). Disruptions to the circadian rhythm, such as damage to the suprachiasmatic nucleus (SCN) in voles (Gerkema and Verhulst 1990) or genetic mutations, like in tau-mutant hamsters (Refinetti 1996), provide a straightforward method to determine if ultradian patterns are a result of circadian organisation and are indeed caused by circadian oscillators. In that case, such patterns ought to vanish or at least change (Refinetti 1996). Otherwise, it must be presumed that independent mechanisms exist (Gerkema and Verhulst 1990). The notion of hyper dian rhythmicity now formally includes circatidal rhythms as well.

Ultradian pacemakers have been localised in numerous instances. This holds true for the pontine oscillator, which regulates REM cycles, as well as spinal pacemakers, which are in charge of rhythms connected to muscular activity. The preoptic area and the arcuate nucleus of the hypothalamus have developed GnRH rhythms, and the retrochiasmatic area is required for the ultradian feeding oscillator in voles (for references, see (Gerkema 2002). Moreover, homeostatic control of ultradian patterns has been shown. Nonetheless, it must be emphasised that it is necessary to define the nature of the underlying ultradian mechanisms in each instance. To differentiate between homeostatic and clock-like control, it is necessary to experimentally modify the periodicity. All of the aforementioned instances exhibit negative feedback as a vital element of this system, but none of them disclose the physiological foundations. Another approach relies on two self-sustaining transcriptional translational oscillations (TTOs), commonly referred to as primary oscillators. These oscillations generate homodimers of transcription-regulating proteins at different ultradian frequencies. Homodimeric transcriptional regulators, consisting of two identical protein molecules, and heterodimeric regulators, made of two distinct protein molecules, are widely recognised (Berger et al. 1999). Additionally, some of these components have been recognised as constituents of various circadian oscillators models and established cellular oscillators (Berger et al. 2002). The oscillation of each fundamental oscillator in the model is regulated by the homodimeric protein products that constitute that oscillator. The activation of a forced oscillator's transcription is initiated by a heterodimeric complex, including one protein molecule from each of the two fundamental oscillators. The behaviour of the forced oscillator exhibits an intricate correlation with the frequency of the basic oscillators. The driven oscillator can exhibit a period that is considerably longer than that of either of the two main oscillators, owing to the specific nature of the interaction between the protein products of the primary oscillators. Numerous models for TTO circadian oscillations that do not include ultradian oscillators as components have also been constructed. One of these (Barkai and Leibler 2000) consists of two genes, one of which generates a transcriptional activator and the other a repressor, both of which have an impact on the other gene and their respective products. In addition, the dimerization of the activator and repressor proteins inactivates both of them.

1.6 The Emergence of a Circadian Neurobiology

How is the circadian system physically regulated in living beings? To find an explanation, we must first understand the neurological underpinnings of the circadian clock. The branch of circadian neurobiology can be categorised into three main areas: (1) the mechanisms and pathways involved in entrainment, (2) the location and function of the pacemaker, and (3) the projections and connections of the pacemaker to effector systems.

1. *Mechanisms and routes of entrainment*: Light is the main stimulus for the circadian system. Previously, it was believed that light did not impact the

synchronisation of human circadian cycles and that social cues were more crucial. This assertion is incorrect. The role of light in humans, as well as in other mammals, is to determine the timing and synchronisation of the circadian pacemaker. This is achieved by particular pathways and biological mechanisms. The impact of light is contingent on time; specifically, light alters the operation of the pacemaker in a time-dependent manner. The phase-response curve (PRC) characterises this phenomenon. A nocturnal creature demonstrates a natural pattern of rest and activity when placed in a situation where it is isolated from normal day-night cycles, such as being in constant darkness or exposed to dim light. The light provided from CT 2 to CT 10 did not impact the natural rhythm of the animal, where the start of its resting time is designated as CT 0 and the beginning of its activity is labelled as CT 12. However, when light is provided between CT 10 and CT 16, the onset of activity is postponed. Light accelerates the start of action from approximately CT 18 to CT 2. The phase delays reflect the pacemaker's perception of light in the CT 10 to CT 16 time period as a lengthening of the day. Light in the CT 18 to CT 2 interval is perceived by the pacemaker as advancing the advent of day; therefore, phase advances reflect this. The full PRC describes the pacemaker's complicated responses to light. Although there is less information regarding human responses to light than about animal responses, current data show that humans exhibit a similar PRC.

2. *Pathways to photic entrainment*: The investigation into the brain mechanisms involved in circadian timing began by examining the visual pathways responsible for regulating entrainment (Moore et al. 1995). Complete severance of all visual pathways originating from the optic chiasm, including all known connections to the central retina, led to blindness in animals. These animals lacked any visual reflexes but maintained a normal circadian rhythm entrainment. This sparked a search for a fresh visual orientation. The discovery was made based on evidence of a direct connection between the retina and the suprachiasmatic nucleus (SCN), which is a small group of cells in the hypothalamus positioned above the optic chiasm and to the side of the third ventricle (RHT; Moore 1973; Moore and Lenn 1972). Since the transection of all other retinal projections does not have any impact on entrainment, it can be concluded that the RHT (retinohypothalamic tract) is sufficient. This is further substantiated by the evidence that partitioning the RHT abolishes entrainment without impacting other visual processes (Johnson et al. 1988). There has been significant focus on the retinal components of the system that controls the synchronisation of biological rhythms with light. The RHT, or Retinal Hypothalamic Tract, consists of a specific group of retinal ganglion cells that are small in size, have few dendritic branches, are evenly spread out across the retina, and are referred to as "W-cell" ganglion cells (Moore et al. 1995). These cells appear to be related distally in the retina to cone bipolar cells and cone photoreceptors, and they solely project to the CTS. According to the research, photoreceptors are not unique to the CTS. Glutamate is the RHT transmitter (Ding et al. 1994; Shirakawa and Moore 1994). Glutamate influences pacemaker activity via NMDA and non-NMDA receptors, most likely via intracellular pathways involving the immediate early

gene, cfos (Rea et al. 1993). The precise methods by which a monophasic stimulus, light, elicits either no response or a biphasic response in SCN neurons are unknown, although new work in Drosophila spp. suggests a possible molecular basis (Myers et al. 1996).

3. *Pathways to non-photic entrainment*: Light is the fundamental circadian Zeitgeber, although it is not the sole stimulus capable of resetting the SCN pacemaker. The retinal ganglion cells that give birth to the RHT innervate a subdivision of the lateral geniculate complex, the intergeniculate leaflet, in addition to the SCN (IGL). The IGL projects back to the SCN via a route that includes GABA and neuropeptide Y (Moore et al. 1995). In vivo (Albers and Ferris 1984) and in vitro (Shibata and Moore 1993) infusions of neuropeptide Y into the SCN, as well as activation of the IGL (Johnson et al. 1988) cause phase alterations with a unique PRC: phase advances during subjective day and phase delays during subjective night. Comparable phase shifts are reported in response to locomotor activity (Reebs and Mrosovsky 1989; Turek 1989), and IGL lesions prevent activity-induced phase alterations. As a result, it appears that the IGL combines photic and non-photic data to deliver entraining information to the SCN pacemaker (Johnson et al. 1989). There are several different afferent inputs to the SCN that are believed to modulate pacemaker activity (Moore 1996). Because locomotor activity is controlled by the clock, its effects on entrainment can be read as pacemaker feedback. Melatonin's action as a pineal hormone is another example of a feedback effect. The pineal gland synthesises and releases melatonin at night in both diurnal and nocturnal species, a rhythm directly controlled by the SCN (Moore 1996) via the routes indicated below. The SCN contains a large number of melatonin receptors in nearly all mammals, including humans (Morgan et al. 1994), and melatonin is the mediator of feedback effects from the pineal gland. Under some settings, melatonin can entrain the SCN pacemaker (Cassone 1990) and a PRC for melatonin has been reported for both animals (Gillette and McArthur 1996) and humans (Lewy et al. 1992).

The pacemaker, or SCN, is circadian neurology's most critical central component. According to four lines of evidence, the SCN is the major mammalian circadian pacemaker. To begin with, the SCN is the principal point of termination of the photic entrainment route, the RHT, and it receives other entraining inputs, as previously mentioned. Second, SCN ablation leads to the loss of practically all circadian rhythms (Meijer and Rietveld 1989; Moore et al. 1995). The only consequence of the SCN lesions in each case is to change the temporal organisation of the function being researched. SCN ablation, for example, eliminates the sleep-wake rhythm while leaving the overall amount of time spent awake and asleep and the fraction of REM and slow-wave sleep unaltered. Finally, even when isolated from the rest of the brain in vivo or in vitro, the SCN exhibits a diurnal rhythm in neuronal activity (Inouye and Kawamura 1979; Shibata and Moore 1993). Fourth, transplanting foetal SCN into the brains of animals rendered arrhythmic by SCN lesions restores rhythmicity, and the restored rhythm follows the donor tissue's intrinsic rhythmicity (Ralph et al. 1990). In mammals, there is evidence that the eye has an independent circadian oscillator (Tosini and Menaker 1996); however, there is no evidence that this affects function outside of the eye. Similarly, while other tissues may have local circadian oscillators, the SCN appears to be the only brain circadian pacemaker that provides temporal organisation for neural adaptation mechanisms.

In all mammals, the SCN is divided into two sections: a core that receives primarily visual input and a shell that gets input from other locations (Cassone et al. 1988). These subdivisions can be distinguished by the separation of afferents as well as the peptide content of the neurons that comprise them. Yet, it appears that all SCN neurons in animals (Moore and Speh 1993) and humans (Gao and Moore 1996) produce GABA, indicating that the output is inhibitory. There are two potential organisational concepts for the SCN pacemaker: Circadian function may evolve as an emergent characteristic via the coupling of a sequence of ultradian oscillators, or individual SCN neurons may be born as circadian oscillators linked by neuronal connections to create a pacemaker. According to recent data, the latter arrangement is in use. When SCN neurons are detached and cultured, they retain individual firing patterns that reflect circadian rhythmicity (Welsh et al. 1995). It is unclear whether all SCN neurons are circadian oscillators, whether this feature is expressed primarily by SCN core neurons, or whether it is represented by another subpopulation. The SCN has a rhythm in neuronal activity, with a high discharge rate during the day and a low discharge rate at night (Inouye and Kawamura 1979; Shibata and Moore 1993). As a result, the signal sent to other brain areas is straightforward. Likewise, the pattern of SCN efferent projections is restricted. The hypothalamus receives the most projections, with the basal forebrain and midline thalamus receiving fewer (Watts 1991). Although these limited projections do not appear to account for the many functions controlled by the clock, studying areas receiving secondary projections (Fig. 1.3) provides some insight. Primary projections to the basal forebrain and midline thalamus transfer information to secondary forebrain areas engaged in a range of higher tasks. Circadian information is sent to the anterior pituitary, hypothalamic, and brainstem reticular formation regions, which are involved in autonomic regulation, metabolism and body temperature management, and the temporal structuring of sleep-wake cycles. Direct SCN projections to the paraventricular hypothalamic nucleus, from there to the upper thoracic inter-mediolateral cell

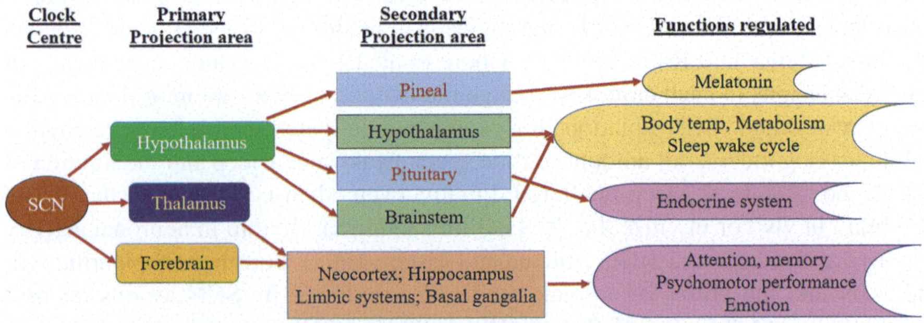

Fig. 1.3 Schematic representation of SCN's projection network

column, and from there to the superior cervical ganglion sympathetic neurons, which project to the pineal, govern pineal activity (Moore 1996).

1.7 Physiological Significance of Circadian Rhythms

To understand the effect of the clock on animal physiology, we must first consider another crucial component of body clocks: peripheral clocks. The influence of the circadian clock on physiology will then be discussed.

1.7.1 The Peripheral Clocks

Preliminary studies suggest the existence of periodic metabolic patterns in cultured heart and adrenal gland samples (Andrews and Edgar Folk 1964). It was shown that rhythms continue to exist in isolated non-neural cultures for many cell divisions, indicating that non-SCN cells possess inherent circadian oscillators. After the core clock genes were cloned, researchers identified rhythmic patterns in the expression of clock mRNA and/or protein in various cells and tissues in the body (Mohawk et al. 2012). Transcriptomic investigations have shown that the regulation of circadian rhythms is very particular to different tissues (Panda et al. 2003; Storch et al. 2002). Furthermore, approximately 50% of all genes that encode proteins in mice and humans are transcribed in a manner that follows a circadian pattern (Ruben et al. 2018). While numerous genes are active in tissues on a daily rhythm, it is important to note that not all of them are considered clock genes. Instead, the phases of the basic clock components in each cell precisely synchronise the daily transcriptional activities of numerous cellular pathways and gene families, which differ significantly depending on the kind of cell. The peripheral clocks provide self-sustaining circadian oscillations in gene expression and organ function. However, the SCN plays a crucial role in coordinating rhythmic behaviour in the periphery (Akhtar et al. 2002). While the core clock machinery remains largely unchanged across organs (Mavroudis et al. 2018), various factors such as circadian changes in hormones like melatonin and cortisol, rhythmic autonomic control, and indirect cues like body temperature or feeding cycles can individually influence the timing and strength of peripheral clock components and consequently affect the patterns of the output pathways they govern. Peripheral clocks play a crucial and unique role in every tissue by controlling the daily expression of certain genes that are involved in several physiological functions, including xenobiotic metabolism, glucose and lipid balance, cellular repair, and inflammation (Lamia et al. 2009). The periodic variation in the outer regions seems to be a vital characteristic of homeostasis, intended to optimise the function of each tissue within a 24-h cycle. The suprachiasmatic nucleus (SCN) is believed to control the peripheral tissue clocks by a mix of peripheral autonomic innervation, endocrine signalling, temperature, and behavioural signals such as food (Mohawk et al. 2012). This text focuses on two fundamental approaches: autonomic nerve control and endocrine signalling. Sympathetic and parasympathetic neurons receive input from the SCN and hypothalamus and innervate the liver, muscle, adipose tissue, gut, heart, and ovary. The sympathetic innervations transmit light/dark signals from the SCN to the pineal gland, which is a small, highly vascularized, secretory neuroendocrine organ positioned on the roof of the third ventricle. The photic information from the SCN influences autonomic nerve activity (Arendt and Aulinas 2000). The suprachiasmatic nucleus (SCN) regulates the secretion of melatonin by synaptic transmission, which involves the release of neurotransmitters such as glutamate, gamma-aminobutyric acid, and norepinephrine (Arendt and Aulinas 2000). Melatonin, produced by the pineal gland, plays a crucial role in regulating sleep-wake cycles and serves as a reliable indicator of the phase of the internal circadian clock (Lewy et al. 1999). Light inhibits melatonin production, causing it to reach its highest levels during the dark period and very low levels during the light period. Melatonin receptors are found in the cells of the suprachiasmatic nucleus (SCN), and melatonin serves as a feedback signal on the SCN. The administration of melatonin from an external source is widely recognised to affect the natural daily rhythm by causing changes in the timing of the internal biological clock (Lewy et al. 1999). Peripheral cells also contain melatonin membrane receptors, which engage with downstream messengers, including adenylyl cyclase, phospholipases, calcium, and potassium channels (Cipolla-Neto and Amaral 2018). The autonomic pathways originating from the suprachiasmatic nucleus (SCN) directly supply the adrenal glands, similar to their innervation of the pineal gland (Fig. 1.4). The splanchnic nerve provides sympathetic innervation that controls the hypothalamic-pituitary-adrenal (HPA) axis. This regulation is responsible for the circadian rhythmicity of peripheral tissues by modulating the generation of epinephrine from the adrenal medulla (Kalsbeek et al. 2012). The release of epinephrine from the sympathetic adrenomedullary system affects the regular release of glucocorticoids by altering the adrenal cortex's responsiveness to adrenocorticotropic hormone (ACTH). The hypothalamic paraventricular nuclei receive photic signals from the SCN and produce corticotropin-releasing hormone (CRH), which stimulates the synthesis of ACTH in the anterior pituitary. In addition to the regulatory influence of epinephrine, ACTH is accountable for the strong circadian rhythm of adrenocortical glucocorticoid hormone secretion. In addition, glucocorticoids that are freely circulating in the body play a role in a rapid negative feedback loop to control the release of CRF from the hypothalamus and ACTH from the pituitary gland (Fig. 1.4). This helps to maintain a balance in the HPA system (Evanson et al. 2010). While melatonin levels regularly increase during periods of darkness and decrease during periods of light, the patterns in the HPA axis are practically reversed in animals active during the day and animals active during the night. The levels of corticosterone reach their highest point in nocturnal animals, such as mice and rats, as they transition from a period of light and inactivity to a period of darkness and activity (Yao et al. 2006). Diurnal species, such as humans, have a rise in cortisol levels during the transition from the dark and inactive period to the light and active period (Challet 2007). Essentially, the circadian cycle of glucocorticoid secretion readies the animal for its next period of activity and food consumption. Therefore, it is crucial to analyse eating patterns and sleep behaviour when

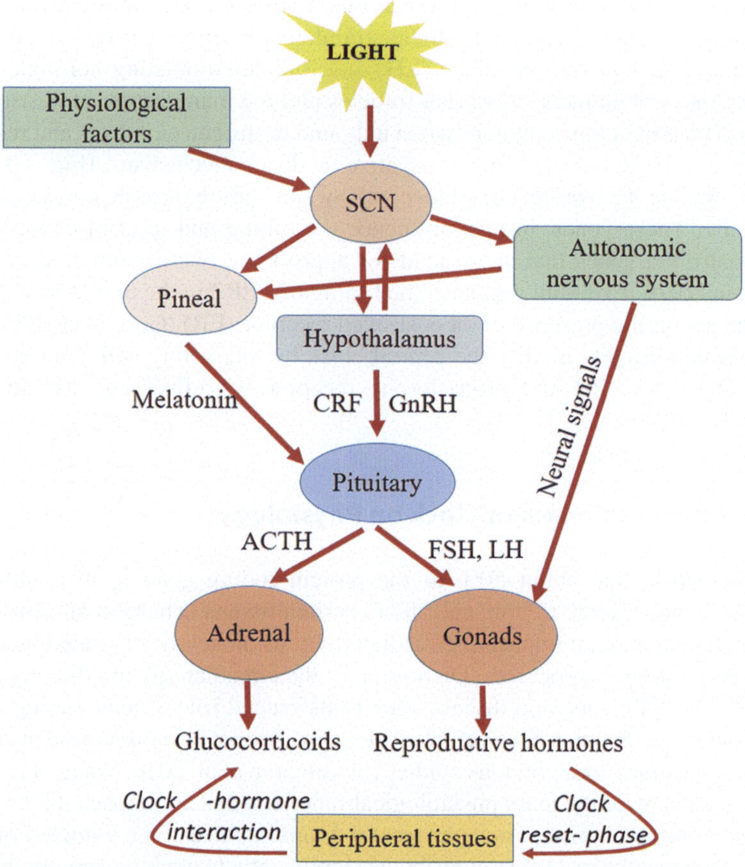

Fig. 1.4 The central and peripheral clock network and their integrative regulation

assessing and comparing circadian trials conducted in animals with human studies. The study conducted by Balsalobre et al. (2000) demonstrated that the administration of the synthetic glucocorticoid and dexamethasone caused temporary changes in the timing of gene expression in peripheral tissues such as the liver, kidney, and heart. This finding provided conclusive proof that glucocorticoids synchronise peripheral biological clocks in living organisms. Recent research has revealed that glucocorticoids have the ability to control the expression of core clock genes (Bmal1, Cry1, Per1, and Per2) as well as numerous clock-controlled genes in almost all tissues. This regulation occurs through the presence of glucocorticoid-response elements (GREs) located in the regulatory regions of these genes (Dickmeis 2009). Both the central and peripheral molecular clocks, such as the hypothalamic-pituitary-adrenal (HPA) axis, have a role in regulating circadian rhythms within the reproductive hypothalamic-pituitary-gonadal (HPG) axis (Fig. 1.4). SCN neurons directly communicate with gonadotropin-releasing hormone (GnRH) neurons in the hypothalamus, exerting an impact on their functioning. GnRH controls the secretion of luteinizing hormone (LH) and follicle-stimulating hormone (FSH) from the anterior pituitary. Luteinizing hormone (LH) and follicle-stimulating hormone (FSH) stimulate the development of ovarian follicles and the manufacture of oestrogen in females. The HPG loop concludes when unbound oestrogen circulates and regulates the release of FSH and LH through negative feedback mechanisms (Fig. 1.4). FSH and LH regulate the ovarian circadian clock by controlling the core clock genes Per and Bmal1. These genes, in turn, influence the timing and level of expression of clock-controlled genes that are crucial for appropriate ovarian function (Sen and Hoffmann 2020). Estradiol regulates the timing of PER2 in the uterus by a process that depends on the presence of an oestrogen receptor (ER) (Gery et al. 2007). Sex hormones can directly modify the central clock by interacting with neurons in the SCN that express ER and progesterone receptors (Kruijver and Swaab 2002) (Fig. 1.4).

1.7.2 Effect of Circadian Clock on Physiology

The observation that about 50% of the protein-coding genome in mammals is expressed in a circadian rhythm specific to certain tissues (Zhang et al. 2014) demonstrates the significant impact of circadian rhythms on nearly all crucial biochemical and physiological processes. Disruption of the circadian rhythm directly affects human health, ailments, and diseases due to its crucial role in maintaining overall bodily and tissue balance. Core-clock genes play a direct or indirect role in regulating these processes, and previous studies (Sukumaran et al. 2010; Wang et al. 2020) offer an overview of the main physiological functions that are impacted by modifying these core-clock genes in experimental animals. Research conducted on individuals experiencing jet lag has contributed significantly to our initial comprehension of the impact of circadian disruption on overall well-being. Time zone changes are commonly recognised to result in minor disturbances in sleep and wake patterns. Studies have shown that frequent travel between different time zones not only leads to temporary cognitive impairments, but also causes anatomical changes in the brain, such as a decrease in the volume of the temporal lobe (Cho 2001). Multiple research on shift workers indicates a correlation between night shifts and a higher incidence of many ailments, such as neurological disorders, diabetes, cancer, and cardiovascular disorders (Woo and Postolache 2008). The widespread use of light-emitting electronic devices, such as smartphones and tablets, can potentially increase the risk of circadian-related disorders for a larger portion of the population, particularly among shift workers who experience disruptions in their sleep-wake cycles (Chang et al. 2015). Various hormones that regulate important physiological processes, such as cortisol, adrenocorticotropic hormone (ACTH), epinephrine, testosterone, Thyroid Stimulating Hormone (TSH), Leptin, and Growth Hormone (GH), are closely controlled by circadian rhythms. These hormones act as a connection between the central and peripheral clocks, as well as their external cues (zeitgebers) (Brandenberger and Weibel 2004; Wittert 2014). Circadian oscillations impact the functioning of physiological systems involved in regulating blood flow and fluid balance, including cardiac output, lymphatic flow, vascular resistance, and glomerular filtration rate. The circadian rhythm regulates the expression of metabolic enzymes and ion channels, ensuring that metabolic and cardiac outputs are optimised. This enables efficient utilisation of nutrients to meet the varying daily demands. Multiple pieces of evidence suggest that circadian rhythms may affect the occurrence of myocardial infarction (Virag and Lust 2014). Several mediators that regulate energy homeostasis activities, such as glycolysis, gluconeogenesis, lipid biosynthesis and metabolism, and oxidative phosphorylation, are under the direct or indirect regulation of core clock genes (Sukumaran et al. 2010; Wang et al. 2020). Disruptions in the timing of these processes can impact energy balance and contribute to conditions such as metabolic syndrome, obesity, and diabetes (Esquirol et al. 2009).

Extensive research spanning several decades has demonstrated a strong correlation between immune-regulatory and inflammatory processes and all significant biological functions. Research indicates that the core clock genes exhibit rhythmic expression in cells responsible for regulating immunological processes, such as innate and adaptive immunity (Boivin et al. 2003). T-lymphocytes, B-lymphocytes, monocytes, neutrophils, and natural killer cells all demonstrate a circadian pattern in their development and movement (Cuesta et al. 2016). Furthermore, a number of significant cytokines, including as TNF-a, interferon-c, IL-6, and IL-2, have substantial diurnal fluctuations (Ray and Reddy 2020). Several inflammatory and autoimmune conditions, such as rheumatoid arthritis, asthma, and systemic lupus, exhibit a notable fluctuation in symptoms and severity during the day. Moreover, the disruption of the host's circadian clock can result in increased pathogen replication and spread, suggesting that circadian rhythms can impact the intensity of acute infections. The intricate connections between various factors have been examined in the context of the COVID-19 pandemic (Morin et al. 2020; Ray and Reddy 2020). A recent publication by Haspel et al. (2020) provides a comprehensive overview of an NIH workshop on circadian rhythms, sleep, and immunity. The publication explores current knowledge and potential future research opportunities to understand the intricate connections between circadian rhythms and immune-modulatory processes. It also highlights the significance of these connections in the development and management of various diseases.

1.8 Health Consequences of Circadian Rhythm Disruption

The circadian system, which regulates numerous rhythms in our bodies, is in high demand in modern times. With the need for goods and services at all hours of the day, increased industrialisation, and economic globalisation, nontraditional work schedules are becoming more frequent. Circadian disturbance is linked to a wide range of significant physiological and psychological diseases, including cardiovascular disease, metabolic syndrome, obesity, immunological dysfunction, an increased risk of cancer, and reproductive issues, according to mounting research.

Moreover, due to self-selection and survival bias, epidemiological studies may understate the negative health consequences of shift work. Ultimately, the consequences of circadian disturbance in human and animal health are enormous. In this section, we present a brief overview of the health effects of circadian disturbance. The effects will be discussed in detail in subsequent chapters of this book.

1.8.1 Neurological Conditions

Circadian disruption is prevalent in neurological disorders, and it has an impact on the progression, presentation, and intensity of the diseases. Circadian disruption has been associated with cerebrovascular disease, epilepsy, migraine, multiple sclerosis, neurodegenerative illnesses, and neurodevelopmental disorders (Abbott et al. 2020). In addition to its association with the severity of sickness, circadian disruption increases the likelihood of developing neurological problems throughout a person's life (Videnovic and Willis 2016). Individuals with autism spectrum disorders (ASDs) and uncommon genetic neurodevelopmental disorders, such as Angelman, Williams, Prader-Willi, fragile X, and Smith-Magenis syndromes, frequently have disruptions in their sleep patterns and have low sleep quality. Autism Spectrum Disorders (ASDs) are common and have been associated with various indicators of disrupted circadian rhythms, including an irregular cortisol pattern and a reduced melatonin pattern (Lorsung et al. 2021). One prevailing hypothesis posits that ASD leads to a deficiency in the production of melatonin. In addition, specific variations in the genes MTNR1A and MTNR1B, known as single-nucleotide polymorphisms (SNPs), have been associated with Autism Spectrum Disorder (ASD) according to a study by Yang et al. (2016). Furthermore, other SNPs in clock genes seem to affect the connection between the severity of ASD and the extent of sleep disruption or circadian rhythm disturbance. Smith-Magenis syndrome (SMS) is a genetic disorder that occurs due to a deletion in a specific region of chromosome 17, known as 17p11.2. This loss includes the retinoic acid-induced 1 (RAI1) gene, which is responsible for the development of the syndrome (Logan and McClung 2019). RAI1 knockout animals had a reduced duration, likely attributed to RAI1 stimulating the transcription of core clock genes. RAI1 plays a role in controlling the release of melatonin in humans (Logan and McClung 2019). Children with SMS experience a reversal of the natural melatonin rhythm and sleep/wake cycle (Lacaria et al. 2013). Moreover, accumulating evidence indicates a two-way connection between disturbances in the circadian rhythm and the development of neurodegenerative conditions, such as traumatic brain injury, Alzheimer's disease (AD), Parkinson's disease (PD), and Huntington's disease. The occurrence of circadian misalignment and reduced magnitude of various rhythms has been associated with the onset of neurodegenerative conditions such as Alzheimer's and Parkinson's (Leng et al. 2020; Lim and Allada 2013; Musiek and Holtzman 2016). Psychiatric diseases, including schizophrenia, depression, bipolar disorder, and seasonal affective disorder, sometimes involve disturbances in the circadian rhythm (Jones and Benca 2015; Walker et al. 2020). Preclinical studies indicate a reciprocal relationship between mood disorders and the circadian system in mice, where alterations in serotonin receptors lead to disturbances in the circadian rhythm (Siemann et al. 2021). Incorrect timing of light exposure in a mouse model can cause depressive symptoms (LeGates and Kvarta 2020). This effect is influenced by specialised retinal cells called intrinsically photosensitive retinal ganglion cells, which send signals to specific brain regions involved in regulating mood, such as the hypothalamus, preoptic area, and amygdala (McGlashan et al. 2021).

1.8.2 Metabolic Dysfunction

Extensive research has been conducted on the impact of circadian disruption on metabolic diseases. The interrelationship between the circadian rhythm and metabolism is crucial for optimising the extraction and utilisation of energy from food during the active period. It also involves digesting stored energy during the rest phase to maintain stable glucose levels throughout the overnight fasting period. Research has demonstrated that circadian dysregulation affects metabolism at various levels, including genetic, cellular, and systemic. This leads to a decrease in glucose tolerance and an increase in insulin resistance, as well as a higher likelihood of developing metabolic syndrome and diabetes (Knutson et al. 2017). The Clock19 mutant mice demonstrate the significant consequences of disrupting the core clock gene, including the development of obesity and metabolic syndrome, which greatly affects metabolism (Turek et al. 2005). The absence of Bmal1 improves the body's response to insulin, as demonstrated by Bass and Takahashi (Bass and Takahashi 2010). Similarly, Maury found in 2019 (Maury 2019) that specific processes of misalignment in tissues also boost insulin sensitivity. Research has shown a connection between social jet lag, which refers to the shifting of bedtimes between work and leisure days, and metabolic syndrome as well as elevated levels of glycosylated haemoglobin (Parsons et al. 2014). The reason why individuals with later chronotypes are more prone to becoming overweight and developing type 2 diabetes could be due to misalignment (Merikanto et al. 2013; Yu et al. 2015). Even among the population of individuals with type 2 diabetes, those who are more inclined towards being active during the nighttime hours exhibited poorer regulation of glucose levels, as evidenced by studies conducted by Iwasaki et al. (2013), Osonoi et al. (2014), and Reutrakul et al. (2013). These studies illustrate that variations in internal and social timing can lead to metabolic dysfunction. Due to the significant correlation between obesity and the Clock19 mutation in mice, it is expected that variations in the CLOCK gene, such as rs6820823, rs3792603, rs11726609 (Riestra et al. 2017), and rs1801260 (Meng et al. 2020), are associated with higher BMI in humans. Identifying CLOCK polymorphisms, such as rs3749474, can enhance the effectiveness of personalised weight reduction treatments, such as dietary fat restriction (Loria-Kohen et al. 2016). Polymorphisms in genes encoding melatonin receptors have been associated with metabolic dysfunction and obesity (Eriksson et al. 2013; Pan et al. 2011). Additionally, these melatonin receptor polymorphisms may partially regulate the relationship between nocturnal food consumption and obesity

(McHill et al. 2017). Reduced daytime energy expenditure following night shifts, along with disturbances in hormonal signalling (leptin/ghrelin) that regulate appetite, are putative mechanisms that connect circadian disruption to weight gain and obesity in individuals working night shifts (Khosravipour et al. 2021).

1.8.3 Cardiovascular Problems

A significant study conducted in 1985 and published in the New England Journal of Medicine examined the temporal distribution of myocardial infarctions (MIs), commonly known as heart attacks. The study revealed a higher likelihood of heart attacks occurring between 6 a.m. and noon (Muller et al. 1985). Further analysis uncovered that this risk arises simultaneously in stroke (Elliott 1998) and ventricular arrhythmias (Twidale et al. 1989). The cause of this delicate morning phase cannot be only attributed to behavioural factors such as eating and sleeping (Edahiro et al. 2014; Scheer et al. 2011). The observed phenomenon seems to be caused, at least partially, by the influence of circadian rhythms on substances that promote blood clotting, such as platelets and prothrombotic plasminogen activator inhibitor-1 (PAI-1; (Pritchett and Reddy 2015; Scheer and Shea 2014). Disruptions in the circadian system have a substantial effect on cardiovascular health due to its role in regulating the timing of circulatory function.

1.8.4 Immune Problems

The circadian modulation of the immune system is a significant area of study that highlights the interconnectedness of the circadian rhythm, the immune system, and sleep (Druzd et al. 2017; Greenberg et al. 2020; Labrecque and Cermakian 2015; Nguyen et al. 2013; Yu et al. 2013). Circadian disruption has been associated with immunological issues. In addition to asthma, certain allergy illnesses, such as anaphylaxis, atopic dermatitis, and allergic rhinitis, have a distinct association with circadian disturbance. It is important to mention that nearly all of these immunological diseases are more common during nighttime. Remarkably, there is a close connection between Covid 19 and the circadian clock (Vitale et al. 2020), and numerous studies are presently underway to explore this relationship.

To summarise, the increasing amount of research that spans from laboratory experiments to clinical applications to population studies highlights the significant impact and interrelatedness of circadian disruption on human health conditions. Additionally, this study underscores the promise of interventions based on circadian rhythms to both prevent and optimise the health of individuals and society as a whole. Furthermore, apart from the effects of circadian disturbance that are particular to certain diseases, there have also been observations of broader effects on other sleep and circadian factors, such as late/evening chronotype and social jet lag, that are applicable across different conditions. Consequently, it is improbable that negative health consequences will happen alone, and many therapeutic approaches like melatonin supplementation or intense light therapy may be used together to tackle multiple issues. Regrettably, there have been limited clinical trials conducted, resulting in therapy suggestions for circadian disorders being relatively ineffective.

The identification of the genetic basis of circadian rhythms, along with the realisation that this genetic machinery is present in almost all cells and tissues and controls metabolic, immune, and neural processes and pathways, has fundamentally transformed our comprehension of the significance of circadian health in medicine. Circadian medicine is an emerging concept that is rapidly gaining popularity worldwide. In addition to translational research, it is crucial to create clinically relevant biomarkers that may be easily applied to a large population. The rapid progress in wearable and environmental sensor technology and the integration of artificial intelligence will allow us to accurately evaluate the timing of various biological rhythms at different levels of organisation in real-time. This will provide a valuable understanding of how the environment, behaviour, and both central and peripheral clocks interact in relation to health and disease. Ultimately, it is necessary to conduct clinical trials and minimise the gap between laboratory research and practical application. For instance, although more than 1000 genes are involved in therapeutic targets or drug metabolism/transport (Zhang et al. 2014), only a small number of commonly prescribed pharmaceuticals have specific recommendations for the time of day they should be taken (Ruben et al. 2019). This is not due to the ineffectiveness of medicines when given at specified times but rather the insufficient nature of clinical studies that do not determine the most suitable timing. Currently, we possess a satisfactory framework for incorporating circadian timing into medical practices, and it is now appropriate to proceed with disease-specific investigations.

References

Abbott SM, Malkani RG, Zee PC (2020) Circadian disruption and human health: a bidirectional relationship. Eur J Neurosci 51(1):567–583. https://doi.org/10.1111/ejn.14298

Akhtar RA, Reddy AB, Maywood ES, Clayton JD, King VM, Smith AG, Gant TW, Hastings MH, Kyriacou CP (2002) Circadian cycling of the mouse liver transcriptome, as revealed by cDNA microarray, is driven by the suprachiasmatic nucleus. Curr Biol 12(7):7

Albers HE, Ferris CF (1984) Neuropeptide Y: role in light-dark cycle entrainment of hamster circadian rhythms. Neurosci Lett 50(1-3):163–168. https://doi.org/10.1016/0304-3940(84)90480-4

Andrews RV, Edgar Folk G (1964) Circadian metabolic patterns in cultured hamster adrenal glands. Comp Biochem Physiol 11(4):393–409. https://doi.org/10.1016/0010-406X(64)90006-4

Arendt J, Aulinas A (2000) Physiology of the pineal gland and melatonin. In: Feingold KR, Anawalt B, Blackman MR, Boyce A, Chrousos G, Corpas E, de Herder WW, Dhatariya K, Dungan K, Hofland J, Kalra S, Kaltsas G, Kapoor N, Koch C, Kopp P, Korbonits M, Kovacs CS, Kuohung W, Laferrère B, Wilson DP (eds) Endotext. MDText.com, Inc, South Dartmouth. http://www.ncbi.nlm.nih.gov/books/NBK550972/.

Balsalobre A, Brown SA, Marcacci L, Tronche F, Kellendonk C, Reichardt HM, Schutz G, Schibler U (2000) Resetting of circadian time in peripheral tissues by glucocorticoid signaling. Science 289:5488

Barkai N, Leibler S (2000) Circadian clocks limited by noise. Nature 403(6767):267–268. https://doi.org/10.1038/35002258

Bass J, Takahashi JS (2010) Circadian integration of metabolism and energetics. Science 330(6009):1349–1354. https://doi.org/10.1126/science.1195027

Berger A, Scheibe K-M, Eichhorn K, Scheibe A, Streich J (1999) Diurnal and ultradian rhythms of behaviour in a mare group of Przewalski horse (Equus ferus przewalskii), measured through one year under semi-reserve conditions. Appl Anim Behav Sci 64(1):1–17. https://doi.org/10.1016/S0168-1591(99)00026-X

Berger A, Scheibe K-M, Brelurut A, Schober F, Streich WJ (2002) Seasonal variation of diurnal and ultradian rhythms in red deer. Biol Rhythm Res 33(3):237–253. https://doi.org/10.1076/brhm.33.3.237.8259

Boivin DB, James FO, Wu A, Cho-Park PF, Xiong H, Sun ZS (2003) Circadian clock genes oscillate in human peripheral blood mononuclear cells. Blood 102(12):4143–4145. https://doi.org/10.1182/blood-2003-03-0779

Brandenberger G, Weibel L (2004) The 24-h growth hormone rhythm in men: sleep and circadian influences questioned. J Sleep Res 13(3):251–255. https://doi.org/10.1111/j.1365-2869.2004.00415.x

Cambrosio A, Keating P (1983) The disciplinary stake: the case of chronobiology. Soc Stud Sci 13(3):323–353. https://doi.org/10.1177/030631283013003001

Campbell FM, Heyes CM, Goldsmith AR (1999) Stimulus learning and response learning by observation in the European starling, in a two-object/two-action test. Anim Behav 58(1):151–158. https://doi.org/10.1006/anbe.1999.1121

Cassone VM (1990) Effects of melatonin on vertebrate circadian systems. Trends Neurosci 13(11):457–464. https://doi.org/10.1016/0166-2236(90)90099-v

Cassone VM, Speh JC, Card JP, Moore RY (1988) Comparative anatomy of the mammalian hypothalamic suprachiasmatic nucleus. J Biol Rhythm 3(1):71–91. https://doi.org/10.1177/074873048800300106

Challet E (2007) Minireview: Entrainment of the suprachiasmatic clockwork in diurnal and nocturnal mammals. Endocrinology 148(12):5648–5655. https://doi.org/10.1210/en.2007-0804

Chang A-M, Aeschbach D, Duffy JF, Czeisler CA (2015) Evening use of light-emitting eReaders negatively affects sleep, circadian timing, and next-morning alertness. Proc Natl Acad Sci 112(4):1232–1237. https://doi.org/10.1073/pnas.1418490112

Cho K (2001) Chronic 'jet lag' produces temporal lobe atrophy and spatial cognitive deficits. Nat Neurosci 4(6):567–568. https://doi.org/10.1038/88384

Cipolla-Neto J, Amaral FGD (2018) Melatonin as a hormone: new physiological and clinical insights. Endocr Rev 39(6):990–1028. https://doi.org/10.1210/er.2018-00084

Cuesta M, Boudreau P, Dubeau-Laramée G, Cermakian N, Boivin DB (2016) Simulated night shift disrupts circadian rhythms of immune functions in humans. J Immunol 196(6):2466–2475. https://doi.org/10.4049/jimmunol.1502422

Dickmeis T (2009) Glucocorticoids and the circadian clock. J Endocrinol 200(1):3–22. https://doi.org/10.1677/JOE-08-0415

Ding JM, Chen D, Weber ET, Faiman LE, Rea MA, Gillette MU (1994) Resetting the biological clock: mediation of nocturnal circadian shifts by glutamate and NO. Science 266(5191):1713–1717. https://doi.org/10.1126/science.7527589

Druzd D, Matveeva O, Ince L, Harrison U, He W, Schmal C, Herzel H, Tsang AH, Kawakami N, Leliavski A, Uhl O, Yao L, Sander LE, Chen C-S, Kraus K, De Juan A, Hergenhan SM, Ehlers M, Koletzko B, Scheiermann C (2017) Lymphocyte circadian clocks control lymph node trafficking and adaptive immune responses. Immunity 46(1):120–132. https://doi.org/10.1016/j.immuni.2016.12.011

Edahiro R, Sakata Y, Nakatani D, Suna S, Usami M, Matsumoto S, Hara M, Kitamura T, Sato H, Yamashita S, Nanto S, Hikoso S, Sakata Y, Hori M, Hamasaki T, Komuro I, OACIS Investigators (2014) Association of lifestyle-related factors with circadian onset patterns of acute myocardial infarction: a prospective observational study in Japan. BMJ Open 4(6):e005067. https://doi.org/10.1136/bmjopen-2014-005067

Elliott WJ (1998) Circadian variation in the timing of stroke onset: a meta-analysis. Stroke 29(5):992–996. https://doi.org/10.1161/01.STR.29.5.992

Emery P, Klarsfeld A, Stanewsky R, Shafer OT (2023) Sensitive timing: a reappraisal of chronobiology's foundational texts. J Biol Rhythm 38(3):245–258. https://doi.org/10.1177/07487304231169080

Eriksson A-K, Van Den Donk M, Hilding A, Östenson C-G (2013) Work stress, sense of coherence, and risk of type 2 diabetes in a prospective study of middle-aged Swedish men and women. Diabetes Care 36(9):2683–2689. https://doi.org/10.2337/dc12-1738

Esquirol Y, Bongard V, Mabile L, Jonnier B, Soulat J, Perret B (2009) Shift work and metabolic syndrome: respective impacts of job strain, physical activity, and dietary rhythms. Chronobiol Int 26(3):544–559. https://doi.org/10.1080/07420520902821176

Evanson NK, Tasker JG, Hill MN, Hillard CJ, Herman JP (2010) Fast feedback inhibition of the HPA axis by glucocorticoids is mediated by endocannabinoid signaling. Endocrinology 151(10):4811–4819. https://doi.org/10.1210/en.2010-0285

Gao B, Moore RY (1996) Glutamic acid decarboxylase message isoforms in human suprachiasmatic nucleus. J Biol Rhythm 11(2):172–179. https://doi.org/10.1177/074873049601100209

Gerkema MP (2002) Ultradian rhythms. In: Kumar V (ed) Biological rhythms. Springer, Berlin, pp 207–215. https://doi.org/10.1007/978-3-662-06085-8_17

Gerkema MP, Verhulst S (1990) Warning against an unseen predator: a functional aspect of synchronous feeding in the common vole, Microtus arvalis. Anim Behav 40(6):1169–1178. https://doi.org/10.1016/S0003-3472(05)80183-6

Gery S, Virk RK, Chumakov K, Yu A, Koeffler HP (2007) The clock gene Per2 links the circadian system to the estrogen receptor. Oncogene 26(57):7916–7920. https://doi.org/10.1038/sj.onc.1210585

Gillette MU, McArthur AJ (1996) Circadian actions of melatonin at the suprachiasmatic nucleus. Behav Brain Res 73(1-2):135–139. https://doi.org/10.1016/0166-4328(96)00085-x

Greenberg EN, Marshall ME, Jin S, Venkatesh S, Dragan M, Tsoi LC, Gudjonsson JE, Nie Q, Takahashi JS, Andersen B (2020) Circadian control of interferon-sensitive gene expression in murine skin. Proc Natl Acad Sci 117(11):5761–5771. https://doi.org/10.1073/pnas.1915773117

Halberg F (1960) Temporal coordination of physiologic function. Cold Spring Harb Symp Quant Biol 25:289–310. https://doi.org/10.1101/SQB.1960.025.01.031

Haspel JA, Anafi R, Brown MK, Cermakian N, Depner C, Desplats P, Gelman AE, Haack M, Jelic S, Kim BS, Laposky AD, Lee YC, Mongodin E, Prather AA, Prendergast B, Reardon C, Shaw AC, Sengupta S, Szentirmai É, Solt LA (2020) Perfect timing: circadian rhythms, sleep, and immunity—an NIH workshop summary. JCI Insight 5(1):e131487. https://doi.org/10.1172/jci.insight.131487

Inouye ST, Kawamura H (1979) Persistence of circadian rhythmicity in a mammalian hypothalamic 'island' containing the suprachiasmatic nucleus. Proc Natl Acad Sci USA 76(11):11. https://doi.org/10.1073/pnas.76.11.5962

Iwasaki M, Hirose T, Mita T, Sato F, Ito C, Yamamoto R, Someya Y, Yoshihara T, Tamura Y, Kanazawa A, Kawamori R, Fujitani Y, Watada H (2013) Morningness–eveningness questionnaire score correlates with glycated hemoglobin in middle-aged male workers with type 2 diabetes mellitus. J Diabetes Invest 4(4):376–381. https://doi.org/10.1111/jdi.12047

Johnson RF, Moore RY, Morin LP (1988) Loss of entrainment and anatomical plasticity after lesions of the hamster retinohypothalamic tract. Brain Res 460(2):297–313. https://doi.org/10.1016/0006-8993(88)90374-5

Johnson RF, Moore RY, Morin LP (1989) Lateral geniculate lesions alter circadian activity rhythms in the hamster. Brain Res Bull 22(2):411–422. https://doi.org/10.1016/0361-9230(89)90068-3

Jones SG, Benca RM (2015) Circadian disruption in psychiatric disorders. Sleep Med Clin 10(4):481–493. https://doi.org/10.1016/j.jsmc.2015.07.004

Kalsbeek A, Van Der Spek R, Lei J, Endert E, Buijs RM, Fliers E (2012) Circadian rhythms in the hypothalamo–pituitary–adrenal (HPA) axis. Mol Cell Endocrinol 349(1):20–29. https://doi.org/10.1016/j.mce.2011.06.042

Khosravipour M, Khanlari P, Khazaie S, Khosravipour H, Khazaie H (2021) A systematic review and meta-analysis of the association between shift work and metabolic syndrome: the roles of sleep, gender, and type of shift work. Sleep Med Rev 57:101427. https://doi.org/10.1016/j.smrv.2021.101427

Knutson KL, Wu D, Patel SR, Loredo JS, Redline S, Cai J, Gallo LC, Mossavar-Rahmani Y, Ramos AR, Teng Y, Daviglus ML, Zee PC (2017) Association between sleep timing, obesity, diabetes: the Hispanic Community Health Study/Study of Latinos (HCHS/SOL) cohort study. Sleep 40(4):14. https://doi.org/10.1093/sleep/zsx014

Kruijver FP, Swaab DF (2002) Sex hormone receptors are present in the human suprachiasmatic nucleus. Neuroendocrinology 75(5):296–305

Labrecque N, Cermakian N (2015) Circadian clocks in the immune system. J Biol Rhythm 30(4):277–290. https://doi.org/10.1177/0748730415577723

Lacaria M, Gu W, Lupski JR (2013) Circadian abnormalities in mouse models of Smith–Magenis syndrome: evidence for involvement of RAI1. Am J Med Genet A 161(7):1561–1568

Lamia KA, Sachdeva UM, DiTacchio L, Williams EC, Alvarez JG, Egan DF, Vasquez DS, Juguilon H, Panda S, Shaw RJ, Thompson CB, Evans RM (2009) AMPK regulates the circadian clock by cryptochrome phosphorylation and degradation. Science 326(5951):437–440. https://doi.org/10.1126/science.1172156

LeGates TA, Kvarta MD (2020) Illuminating a path from light to depression. Nat Neurosci 23(7):785–787. https://doi.org/10.1038/s41593-020-0659-x

Leng Y, Blackwell T, Cawthon PM, Ancoli-Israel S, Stone KL, Yaffe K (2020) Association of circadian abnormalities in older adults with an increased risk of developing Parkinson disease. JAMA Neurol 77(10):1270–1278

Lewy AJ, Ahmed S, Jackson JM, Sack RL (1992) Melatonin shifts human circadian rhythms according to a phase-response curve. Chronobiol Int 9(5):380–392. https://doi.org/10.3109/07420529209064550

Lewy AJ, Cutler NL, Sack RL (1999) The endogenous melatonin profile as a marker for circadian phase position. J Biol Rhythm 14(3):227–236. https://doi.org/10.1177/074873099129000641

Lim C, Allada R (2013) Emerging roles for post-transcriptional regulation in circadian clocks. Nat Neurosci 16(11):11

Logan RW, McClung CA (2019) Rhythms of life: circadian disruption and brain disorders across the lifespan. Nat Rev Neurosci 20(1):49–65. https://doi.org/10.1038/s41583-018-0088-y

Loria-Kohen V, Espinosa-Salinas I, Marcos-Pasero H, Lourenço-Nogueira T, Herranz J, Molina S, Reglero G, Ramirez De Molina A (2016) Polymorphism in the CLOCK gene may influence the effect of fat intake reduction on weight loss. Nutrition 32(4):453–460. https://doi.org/10.1016/j.nut.2015.10.013

Lorsung E, Karthikeyan R, Cao R (2021) Biological timing and neurodevelopmental disorders: a role for circadian dysfunction in autism spectrum disorders. Front Neurosci 15:642745

Maury E (2019) Off the clock: from circadian disruption to metabolic disease. Int J Mol Sci 20(7):1597. https://doi.org/10.3390/ijms20071597

Mavroudis PD, DuBois DC, Almon RR, Jusko WJ (2018) Daily variation of gene expression in diverse rat tissues. PLoS ONE 13(5):e0197258. https://doi.org/10.1371/journal.pone.0197258

McGlashan EM, Poudel GR, Jamadar SD, Phillips AJK, Cain SW (2021) Afraid of the dark: light acutely suppresses activity in the human amygdala. PLoS ONE 16(6):e0252350. https://doi.org/10.1371/journal.pone.0252350

McHill AW, Phillips AJ, Czeisler CA, Keating L, Yee K, Barger LK, Garaulet M, Scheer FA, Klerman EB (2017) Later circadian timing of food intake is associated with increased body fat. Am J Clin Nutr 106(5):1213–1219. https://doi.org/10.3945/ajcn.117.161588

Meijer JH, Rietveld WJ (1989) Neurophysiology of the suprachiasmatic circadian pacemaker in rodents. Physiol Rev 69(3):671–707. https://doi.org/10.1152/physrev.1989.69.3.671

Meng Y, Lohse B, Cunningham-Sabo L (2020) Sex modifies the association between the CLOCK variant rs1801260 and BMI in school-age children. PLoS ONE 15(8):e0236991. https://doi.org/10.1371/journal.pone.0236991

Merikanto I, Lahti T, Puolijoki H, Vanhala M, Peltonen M, Laatikainen T, Vartiainen E, Salomaa V, Kronholm E, Partonen T (2013) Associations of chronotype and sleep with cardiovascular

diseases and type 2 diabetes. Chronobiol Int 30(4):470–477. https://doi.org/10.3109/07420528.2012.741171
Mohawk JA, Green CB, Takahashi JS (2012) Central and peripheral circadian clocks in mammals. Annu Rev Neurosci 35(1):445–462. https://doi.org/10.1146/annurev-neuro-060909-153128
Moore RY (1973) Retinohypothalamic projection in mammals: a comparative study. Brain Res 49(2):403–409. https://doi.org/10.1016/0006-8993(73)90431-9
Moore RY (1996) Neural control of the pineal gland. Behav Brain Res 73(1–2):125–130. https://doi.org/10.1016/0166-4328(96)00083-6
Moore RY, Lenn NJ (1972) A retinohypothalamic projection in the rat. J Comp Neurol 146(1):1–14. https://doi.org/10.1002/cne.901460102
Moore RY, Speh JC (1993) GABA is the principal neurotransmitter of the circadian system. Neurosci Lett 150(1):112–116. https://doi.org/10.1016/0304-3940(93)90120-a
Moore RY, Speh JC, Card JP (1995) The retinohypothalamic tract originates from a distinct subset of retinal ganglion cells. J Comp Neurol 352(3):351–366. https://doi.org/10.1002/cne.903520304
Morgan PJ, Barrett P, Howell HE, Helliwell R (1994) Melatonin receptors: localization, molecular pharmacology and physiological significance. Neurochem Int 24(2):101–146. https://doi.org/10.1016/0197-0186(94)90100-7
Morin CM, Carrier J, Bastien C, Godbout R, Canadian Sleep and Circadian Network (2020) Sleep and circadian rhythm in response to the COVID-19 pandemic. Can J Public Health 111(5):654–657. https://doi.org/10.17269/s41997-020-00382-7
Muller JE, Stone PH, Turi ZG, Rutherford JD, Czeisler CA, Parker C, Poole WK, Passamani E, Roberts R, Robertson T, Sobel BE, Willerson JT, Braunwald E (1985) Circadian variation in the frequency of onset of acute myocardial infarction. N Engl J Med 313(21):1315–1322. https://doi.org/10.1056/NEJM198511213132103
Musiek ES, Holtzman DM (2016) Mechanisms linking circadian clocks, sleep, and neurodegeneration. Science 354(6315):1004–1008. https://doi.org/10.1126/science.aah4968
Myers MP, Wager-Smith K, Rothenfluh-Hilfiker A, Young MW (1996) Light-induced degradation of TIMELESS and entrainment of the Drosophila circadian clock. Science 271(5256):1736–1740. https://doi.org/10.1126/science.271.5256.1736
Nguyen KD, Fentress SJ, Qiu Y, Yun K, Cox JS, Chawla A (2013) Circadian gene Bmal1 regulates diurnal oscillations of Ly6C ^{hi} inflammatory monocytes. Science 341(6153):1483–1488. https://doi.org/10.1126/science.1240636
Osonoi Y, Mita T, Osonoi T, Saito M, Tamasawa A, Nakayama S, Someya Y, Ishida H, Kanazawa A, Gosho M, Fujitani Y, Watada H (2014) Morningness–eveningness questionnaire score and metabolic parameters in patients with type 2 diabetes mellitus. Chronobiol Int 31(9):1017–1023. https://doi.org/10.3109/07420528.2014.943843
Pan A, Schernhammer ES, Sun Q, Hu FB (2011) Rotating night shift work and risk of type 2 diabetes: two prospective cohort studies in women. PLoS Med 8(12):e1001141. https://doi.org/10.1371/journal.pmed.1001141
Panda S, Provencio I, Tu DC, Pires SS, Rollag MD, Castrucci AM, Pletcher MT, Sato TK, Wiltshire T, Andahazy M (2003) Melanopsin is required for non-image-forming photic responses in blind mice. Science 301:5632
Parsons S, Valdez-Ramirez V, Congdon R, Williams S (2014) Contrasting patterns of litterfall seasonality and seasonal changes in litter decomposability in a tropical rainforest region. Biogeosciences 11:18
Pritchett D, Reddy AB (2015) Circadian clocks in the hematologic system. J Biol Rhythm 30(5):374–388. https://doi.org/10.1177/0748730415592729
Ralph MR, Foster RG, Davis FC, Menaker M (1990) Transplanted suprachiasmatic nucleus determines circadian period. Science 247(4945):4945. https://doi.org/10.1126/science.2305266
Ray S, Reddy AB (2020) COVID-19 management in light of the circadian clock. Nat Rev Mol Cell Biol 21(9):494–495. https://doi.org/10.1038/s41580-020-0275-3

Rea MA, Buckley B, Lutton LM (1993) Local administration of EAA antagonists blocks light-induced phase shifts and c-fos expression in hamster SCN. Am J Phys 265(5):1191–1198. https://doi.org/10.1152/ajpregu.1993.265.5.R1191

Reebs SG, Mrosovsky N (1989) Large phase-shifts of circadian rhythms caused by induced running in a re-entrainment paradigm: the role of pulse duration and light. J Comp Physiol A 165(6):819–825

Refinetti R (1996) Comparison of the body temperature rhythms of diurnal and nocturnal rodents. J Exp Zool 275(1):67–70. https://doi.org/10.1002/(SICI)1097-010X(19960501)275:1<67::AID-JEZ10>3.0.CO;2-U

Reutrakul S, Hood MM, Crowley SJ, Morgan MK, Teodori M, Knutson KL, Van Cauter E (2013) Chronotype is independently associated with glycemic control in type 2 diabetes. Diabetes Care 36(9):2523–2529. https://doi.org/10.2337/dc12-2697

Riestra P, Gebreab SY, Xu R, Khan RJ, Gaye A, Correa A, Min N, Sims M, Davis SK (2017) Circadian CLOCK gene polymorphisms in relation to sleep patterns and obesity in African Americans: findings from the Jackson heart study. BMC Genet 18(1):58. https://doi.org/10.1186/s12863-017-0522-6

Ruben MD, Wu G, Smith DF, Schmidt RE, Francey LJ, Lee YY, Anafi RC, Hogenesch JB (2018) A database of tissue-specific rhythmically expressed human genes has potential applications in circadian medicine. Sci Transl Med 10(458):eaat8806. https://doi.org/10.1126/scitranslmed.aat8806

Ruben MD, Smith DF, FitzGerald GA, Hogenesch JB (2019) Dosing time matters. Science 365(6453):547–549. https://doi.org/10.1126/science.aax7621

Scheer FAJL, Shea SA (2014) Human circadian system causes a morning peak in prothrombotic plasminogen activator inhibitor-1 (PAI-1) independent of the sleep/wake cycle. Blood 123(4):590–593. https://doi.org/10.1182/blood-2013-07-517060

Scheer FAJL, Michelson AD, Frelinger AL, Evoniuk H, Kelly EE, McCarthy M, Doamekpor LA, Barnard MR, Shea SA (2011) The human endogenous circadian system causes greatest platelet activation during the biological morning independent of behaviors. PLoS ONE 6(9):e24549. https://doi.org/10.1371/journal.pone.0024549

Schwartz WJ, Daan S (2017) Origins: a brief account of the ancestry of circadian biology. In: Kumar V (ed) Biological timekeeping: clocks, rhythms and behaviour. Springer, New Delhi, pp 3–22. https://doi.org/10.1007/978-81-322-3688-7_1

Sen A, Hoffmann HM (2020) Role of core circadian clock genes in hormone release and target tissue sensitivity in the reproductive axis. Mol Cell Endocrinol 501:110655. https://doi.org/10.1016/j.mce.2019.110655

Shibata S, Moore RY (1993) Neuropeptide Y and optic chiasm stimulation affect suprachiasmatic nucleus circadian function in vitro. Brain Res 615(1):95–100. https://doi.org/10.1016/0006-8993(93)91118-c

Shirakawa T, Moore RY (1994) Glutamate shifts the phase of the circadian neuronal firing rhythm in the rat suprachiasmatic nucleus in vitro. Neurosci Lett 178(1):47–50. https://doi.org/10.1016/0304-3940(94)90286-0

Siemann JK, Grueter BA, McMahon DG (2021) Rhythms, reward, and blues: consequences of circadian photoperiod on affective and reward circuit function. Neuroscience 457:220–234. https://doi.org/10.1016/j.neuroscience.2020.12.010

Storch K-F, Lipan O, Leykin I, Viswanathan N, Davis FC, Wong WH, Weitz CJ (2002) Extensive and divergent circadian gene expression in liver and heart. Nature 417(6884):78–83. https://doi.org/10.1038/nature744

Sukumaran S, Almon RR, DuBois DC, Jusko WJ (2010) Circadian rhythms in gene expression: relationship to physiology, disease, drug disposition and drug action. Adv Drug Deliv Rev 62(9–10):904–917. https://doi.org/10.1016/j.addr.2010.05.009

Tosini G, Menaker M (1996) Circadian rhythms in cultured mammalian retina. Science 272(5260):419–421. https://doi.org/10.1126/science.272.5260.419

Turek FW (1989) Effects of stimulated physical activity on the circadian pacemaker of vertebrates. J Biol Rhythm 4(2):135–147

Turek FW, Joshu C, Kohsaka A, Lin E, Ivanova G, McDearmon E, Laposky A, Losee-Olson S, Easton A, Jensen DR, Eckel RH, Takahashi JS, Bass J (2005) Obesity and metabolic syndrome in circadian clock mutant mice. Science 308(5724):1043–1045. https://doi.org/10.1126/science.1108750

Twidale N, Taylor S, Heddle WF, Ayres BF, Tonkin AM (1989) Morning increase in the time of onset of sustained ventricular tachycardia. Am J Cardiol 64(18):1204–1206. https://doi.org/10.1016/0002-9149(89)90881-3

Videnovic A, Willis GL (2016) Circadian system—a novel diagnostic and therapeutic target in Parkinson's disease? Mov Disord 31(3):260–269

Virag JA, Lust RM (2014) Circadian influences on myocardial infarction. Front Physiol 5:116760

Vitale JA, Perazzo P, Silingardi M, Biffi M, Banfi G, Negrini F (2020) Is disruption of sleep quality a consequence of severe Covid-19 infection? A case-series examination. Chronobiol Int 37(7):1110–1114. https://doi.org/10.1080/07420528.2020.1775241

Walker WH, Walton JC, DeVries AC, Nelson RJ (2020) Circadian rhythm disruption and mental health. Transl Psychiatry 10(1):28. https://doi.org/10.1038/s41398-020-0694-0

Wang H, Yang Z, Li X, Huang D, Yu S, He J, Li Y, Yan J (2020) Single-cell in vivo imaging of cellular circadian oscillators in zebrafish. PLoS Biol 18(3):e3000435. https://doi.org/10.1371/journal.pbio.3000435

Watts AG (1991) Suprachiasmatic nucleus: the mind's clock. J Anat 184(2):431

Welsh DK, Logothetis DE, Meister M, Reppert SM (1995) Individual neurons dissociated from rat suprachiasmatic nucleus express independently phased circadian firing rhythms. Neuron 14(4):697–706. https://doi.org/10.1016/0896-6273(95)90214-7

Wittert G (2014) The relationship between sleep disorders and testosterone in men. Asian J Androl 16(2):262. https://doi.org/10.4103/1008-682X.122586

Woo J-M, Postolache TT (2008) The impact of work environment on mood disorders and suicide: evidence and implications. Int J Disability Human Dev 7(2):185. https://doi.org/10.1515/IJDHD.2008.7.2.185

Yang G, Chen L, Grant GR, Paschos G, Song W-L, Musiek ES, Lee V, McLoughlin SC, Grosser T, Cotsarelis G (2016) Timing of expression of the core clock gene Bmal1 influences its effects on aging and survival. Sci Transl Med 8(324):324

Yao Z, DuBois DC, Almon RR, Jusko WJ (2006) Modeling circadian rhythms of glucocorticoid receptor and glutamine synthetase expression in rat skeletal muscle. Pharm Res 23(4):670–679. https://doi.org/10.1007/s11095-005-9608-3

Yu X, Rollins D, Ruhn KA, Stubblefield JJ, Green CB, Kashiwada M, Rothman PB, Takahashi JS, Hooper LV (2013) T_H 17 cell differentiation is regulated by the circadian clock. Science 342(6159):727–730. https://doi.org/10.1126/science.1243884

Yu JH, Yun C-H, Ahn JH, Suh S, Cho HJ, Lee SK, Yoo HJ, Seo JA, Kim SG, Choi KM, Baik SH, Choi DS, Shin C, Kim NH (2015) Evening chronotype is associated with metabolic disorders and body composition in middle-aged adults. J Clin Endocrinol Metabol 100(4):1494–1502. https://doi.org/10.1210/jc.2014-3754

Zhang R, Lahens NF, Ballance HI, Hughes ME, Hogenesch JB (2014) A circadian gene expression atlas in mammals: implications for biology and medicine. Proc Natl Acad Sci 111(45):16219–16224

Mechanisms of Circadian Oscillations 2

2.1 Introduction

Homeostasis, which refers to the ability of organisms to maintain variables within feasible bounds despite an individual's lifestyle and environment, is a key aspect of physiology. However, when a variable is consistently observed over a 24-h period in individuals who follow a normal daily routine (eating and being active during the day, sleeping at night), circadian patterns emerge. While the specific characteristics of these rhythms may differ on a daily basis, the overall pattern remains consistent among both individuals and among individuals.

Chronobiology examines how an individual's biological processes are influenced by factors such as the time of day, the seasons, or other external influences that operate on different time scales. Two prominent factors contributing to these rhythms include an individual's surroundings and lifestyle. The environment exhibits noticeable variations between day and night, such as changes in light intensity and temperature. Additionally, an individual's lifestyle typically consists of daytime activities (including physical, mental, and social engagement, as well as food and fluid consumption) and nighttime rest and sleep.

The organism's timekeeping mechanism, often known as its biological "clock", enables it to anticipate and adapt to the diurnal variations in the physical environment. This ensures that the organism performs appropriate actions at the appropriate times of the day. The biological clock serves to internally regulate and synchronize internal processes. The alignment between an organism and its external and internal environments is crucial for the organism's health and survival. A lack of alignment between the organism and the external environment can result in detrimental consequences for the individual. For instance, if a nocturnal rodent were to leave its underground dwelling during the daytime, it would be extremely vulnerable to predation by other animals. Likewise, an absence of coordination within the internal environment can result in health issues for the person, such as those linked to jet lag, shift work, and the resulting sleep deprivation (for example, reduced cognitive

S. I. Rizvi, G. Majumdar, *Circadian Clock and Aging*,
https://doi.org/10.1007/978-981-97-6695-6_2

ability, disrupted hormone regulation, and digestive disorders). Researchers in the discipline of chronobiology investigate the processes that govern biological timekeeping systems and the potential repercussions that may arise from their malfunction (Aschoff 1960). Chronobiology, in its most comprehensive sense, encompasses all research domains that investigate biological timing, including rapid cycles (such as hormone secretion occurring in distinct pulses throughout the day), daily cycles (such as activity and rest cycles), and longer cycles that occur monthly or annually (such as reproductive cycles in certain species).

An essential characteristic of circadian rhythms is their inherent ability to sustain themselves. Most daily patterns that occur in natural settings persist in laboratory conditions without any external time signals from the physical environment, such as constant light or darkness. Circadian rhythms that manifest without external environmental cues lasting over a 24-h period are called free running. This implies that the rhythm is not coordinated by any periodic alteration in the physical surroundings. Technically, a diurnal rhythm should not be labelled as circadian until it has been proven to continue under consistent environmental conditions, thus differentiating it from rhythms that only react to 24-h environmental fluctuations. From a practical standpoint, there is minimal need to differentiate between diurnal and circadian rhythms, as nearly all daily rhythms are actually circadian. Furthermore, there is no differentiation in nomenclature regarding circadian rhythms based on the specific environmental stimulation that aligns the cycle.

The continued presence of rhythms in the absence of a dark-light cycle or any external time signal, such as a *Zeitgeber*, strongly suggests the presence of an internal timekeeping device, sometimes known as a biological clock. Nevertheless, certain researchers have highlighted that the continued presence of rhythmic patterns does not automatically rule out the potential influence of further unregulated cycles caused by the Earth's rotation on its axis.

The possibility that unregulated geomagnetic cues may contribute to the endurance of rhythmicity can be disproven by another distinctive aspect of circadian rhythms: These cycles endure approximately, but not precisely, 24 h. If the rhythms were externally stimulated, they would continue with a precise 24-h period. The seeming lack of precision is a significant characteristic of rhythmicity, nonetheless. Pittendrigh (1960) showed that the departure from a 24-h cycle serves as a mechanism for the internal timekeeping system to constantly synchronise with the light-dark environment. The continual adjustment ensures enhanced precision in regulating the timing, or phase, of the stated rhythms, as minimal deviation is permitted before the rhythm is reset to the accurate phase.

One additional defining feature of circadian rhythms is their capacity to be synchronized, or entrained, by external temporal signals, such as the alternating periods of light and darkness. Therefore, while circadian rhythms can continue without external time cues (indicating that they are not influenced by the surroundings), often these cues are present and the rhythms are synchronized with them. Hence, in the event of a change in external stimuli (such as travelling across different time zones), the biological rhythms will adjust to synchronize with the new stimuli. The term used to describe this alignment is known as entrainment. Details of the characteristics and parameters of circadian rhythms have been depicted in Chap. 1.

At first, it was uncertain whether entrainment was accomplished by adjusting the rate of cycling (i.e., by either shortening or lengthening the cycle until it matched the new cues and then returning to its normal length) or by discrete "resetting" events. The disagreement prompted experiments that yielded significant breakthroughs. Researchers found that the organism's reaction to light, whether the rhythm cycle advances, delays, or remains the same, varies depending on the phase of the cycle to which it is exposed (Pittendrigh 1960). Consequently, when an individual is exposed to light at the beginning of their usual dark period, it typically leads to a delay in their biological rhythm. Conversely, exposure to light towards the end of their usual dark period usually causes an advancement in their biological rhythm. The variation in responses can be depicted using a phase-response curve (see Fig. 2.1). This curve can forecast how an organism will adjust to changes in both the regular light-dark cycles and atypical light cycles, such as cycles that are not 24 h long or have varying ratios of light to darkness. The presence of a phase-response curve indicates that entrainment is accomplished by discrete resetting events rather from alterations in the cycling rate. The intensity of light can also influence the duration of cycle periods in organisms that are exposed to steady light, in addition to the time of the light exposure. Hence, increased exposure to higher light intensities can extend the duration in certain species and reduce it in others. This phenomenon has been named "Aschoff's rule" (Aschoff 1960). Ultimately, both modes of entrainment seem to be different facets of the same phenomenon, as the outcomes of Aschoff's rule can be anticipated or elucidated by the phase-response curves to light.

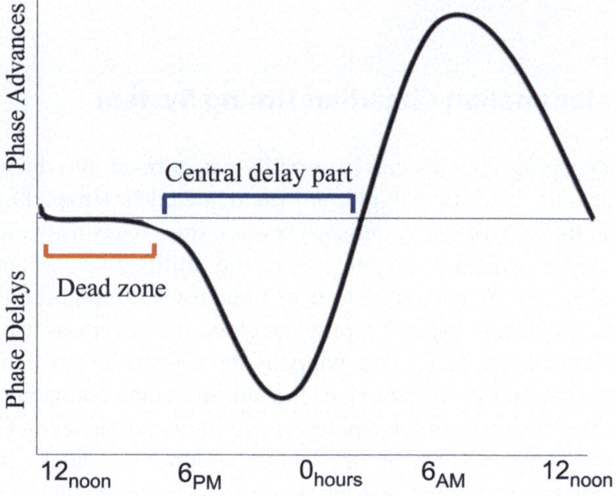

Fig. 2.1 Representation of a typical phase response curve

While the light-dark cycle is undoubtedly the primary external cue that regulates the biological rhythms of all creatures, other elements, such as social interactions, physical activity, and even temperature, can also influence the timing of these rhythms. The impact of temperature on circadian rhythms is particularly intriguing since a shift in temperature can modify the phase of a cycle without significantly changing the frequency of cycling. This indicates that the cycle may commence earlier or later than usual but maintain the same duration. On one side, the pacemaker of the internal clock's ability to compensate for temperature fluctuations is crucial for its capacity to anticipate and adjust to environmental changes. This is because a clock that fluctuates its speed in response to temperature changes would be ineffective. However, temperature adjustment can be perplexing due to the fact that several biological processes, such as biochemical reactions in the body, are influenced by temperature alterations, either speeding up or slowing down. In essence, this conundrum has revealed a hint about the inherent characteristics of the internal clock, specifically that circadian rhythms are genetically determined. A pattern of gene expression would exhibit more resilience to changes in temperature compared to a basic biochemical activity. Two ultimate characteristics of circadian rhythms also offer significant indications of the composition of the rhythms. One notable characteristic of these rhythms is their widespread occurrence in nature: Circadian rhythms are present in a wide range of biological processes and creatures, exhibiting similar characteristics and even displaying similar responses to light. Circadian rhythms are believed to originate at the cellular level, as evidenced by the similarity between the rhythms of unicellular species (such as algae or the dinoflagellate *Gonyaulax*) and those of more complex humans. Both of these discoveries indicate that a periodic pattern in the activation (i.e. expression) of specific genes may be responsible for the mechanism that keeps track of time.

In this chapter, we will focus on the 'circadian clock' as studied in mammals and try to explain its basic mechanistic working both at the anatomical and molecular levels.

2.2 The Mammalian Circadian Timing System

Biological timekeeping systems can be broadly categorised into three components (Eskin 1979) as shown in the eskinogram (as depicted in Chap. 1). A biological clock, located at the core of the system, serves as a time-measuring mechanism. An effective timekeeping mechanism must have the ability to be changed or reset. Therefore, input routes are necessary to detect and transmit important environmental stimuli to the clock. Likewise, a biological clock is not very useful if it does not activate physiological reactions. This happens through pathways that result in the visible patterns of rhythms we perceive. Indeed, the connections among different components of the circadian timekeeping system in mammals are not one-way (for example, overt rhythms can have an impact on the biological clock), and the ability for biological timekeeping is not restricted to a single anatomical structure.

The regulation of circadian rhythms in mammals is governed by a central circadian pacemaker located in the suprachiasmatic nuclei (SCN) (Mohawk and Takahashi 2011; Weaver and Emery 2013). The suprachiasmatic nucleus possesses distinct characteristics that set it apart. It receives crucial input from the retina, which is essential for synchronising with light. Additionally, it has the ability to transmit information about the time of day to other brain structures and, consequently, to other tissues. Contrary to the prevailing dogmatic belief until the late 1990s, it is now known that the SCN are not the only ones capable of molecular oscillations with a cycle period of around 24 h (Balsalobre et al. 1998). The genes that are essential for the regulation of circadian rhythms are expressed in various tissues, and many peripheral tissues exhibit rhythmic gene expression both in vivo and in vitro (when removed and cultured (Abe et al. 2002; Menaker et al. 2013; Yamazaki et al. 2000)). Given that a majority of cells possess the ability to undergo cell-autonomous circadian oscillations, what is the reason for the exceptional status of the SCN in the field of circadian research? The solution can be found in the differentiation between an oscillator and a pacemaker. SCN neurons engage in intercellular communication, resulting in the establishment of a rhythmic pattern at the tissue level, which can endure when studied in a laboratory setting. SCN explants have the ability to undergo sustained oscillations for extended periods of time, ranging from months to even years, when cultured in vitro (refer to Yamazaki et al. 2000). However, the suprachiasmatic nucleus performs functions beyond oscillation. The suprachiasmatic nucleus transmits its rhythmic patterns through neuroanatomical pathways, resulting in physiological and behavioural rhythms. These rhythms synchronise the activities of many peripheral tissues and act as a pacemaker. The circadian timing system in mammals operates in a hierarchical manner. Non-SCN "peripheral oscillators" are synchronised by various rhythms regulated by the SCN, including body temperature, corticosteroid levels, melatonin levels, food intake, and direct autonomic innervation (Balsalobre et al. 2000; Gerber et al. 2013; Wood and Loudon 2014). Blood-borne signals synchronise peripheral tissues, partially through the involvement of serum response factor-1 (Gerber et al. 2013). Additionally, the liver oscillator's entrainment to meals is mediated by poly (ADP-ribose) polymerase 1(Asher et al. 2010). Without the suprachiasmatic nucleus (SCN), both behavioural rhythms and most physiological rhythms cease to exist. However, evidence suggests the continuation of certain oscillations at the tissue level (Saini et al. 2013).

Hence, the SCN plays a crucial role in maintaining internal temporal order by coordinating the activity of circadian oscillators in peripheral tissues (Menaker et al. 2013; Saini et al. 2013).

Below in this section, we will describe the "input" and "pacemaker" of the mammalian circadian system.

2.2.1 Input to the Clock: Retina and Non-retinal Origins

Multiple investigations have confirmed that the retina is the sole tissue in mammals that can perceive light and non-light stimuli (Yamazaki et al. 2000). Hence, the retina is crucial in regulating the circadian rhythm of the entire organism, as it is the sole provider of light-related information to the SCN and, consequently, to the rest of the body. There is a suggestion that the retina may play a role in the general regulation of the circadian rhythm, as alterations in the period of the circadian rhythm in locomotor activity have been noticed following the degeneration of photoreceptors or the removal of both eyes (Tosini et al. 2007). Nevertheless, the duration of free-running activity in rd/rd mice and mice without any of the recognised retinal photopigments (such as rhodopsin, cone opsins, and melanopsin) is indistinguishable from that in wild-type mice (Panda et al. 2003). Collectively, these data suggest that forecasting the impact of the retinal circadian clock system on the free-running period may pose challenges. The causality of the observed alterations remains uncertain, as it is unclear whether they are attributed to interactions between the two clock systems or to other unidentified mechanisms that could be influenced by photoreceptor degradation or removal of the eyes. Nevertheless, the retina's role in transmitting light information to the central clock in the SCN is definite.

The primary input to the suprachiasmatic nucleus (SCN) originates in the retina. The ipRGCs, which express melanopsin, send their projections to the suprachiasmatic nucleus (SCN) and other brain regions involved in non-image-forming vision, such as the olivary pretectal nucleus (OPN) and intergeniculate leaflet (IGL) (Baver et al. 2008; Morin 2012; Paul et al. 2012). Previously, it was believed that ipRGCs mainly connect with neurons that produce VIP and GRP in the ventromedial SCN core because to their specific response to light stimulation (Hughes et al. 2004). Nevertheless, studies have demonstrated that individual intrinsically photosensitive retinal ganglion cells (ipRGCs) innervate neurons in both the central and dorsolateral regions of the suprachiasmatic nucleus, including those responsible for producing arginine vasopressin (AVP) (Fernandez et al. 2016). Light pulses that reset circadian cycles cause intrinsically photosensitive retinal ganglion cells (ipRGCs) to release both the excitatory neurotransmitter glutamate and the modulatory neuropeptide pituitary adenylate cyclase-activating polypeptide (PACAP) onto SCN neurons (Webb et al. 2013). Remarkably, a specific group of intrinsically photosensitive retinal ganglion cells (ipRGCs) release gamma-aminobutyric acid (GABA) into the suprachiasmatic nucleus to reduce the responsiveness of the circadian system to low-intensity light stimuli (Sonoda et al. 2020). The activation threshold for neurons in the dorsolateral SCN is likely larger than that for neurons in the ventromedial SCN when it comes to ipRGC-mediated signals. Activation of glutamate and PACAP receptors leads to an increase in calcium influx into core SCN neurons. This occurs through the opening of calcium-permeable N-methyl-D-aspartate (NMDA) glutamate receptors and by raising the firing rate, both directly and indirectly. The reaction is subsequently conveyed intracellularly to the molecular clock by activating calcium-dependent kinases and intercellularly to other SCN neurons by releasing VIP or GABA (Quintero et al. 2003). The projections of intrinsically photosensitive retinal ganglion cells (ipRGCs) to the suprachiasmatic nucleus (SCN) through the retino-hypothalamic tract (RHT), which were initially discovered in nocturnal mice, are mostly preserved in diurnal species (Langel et al. 2015). Nevertheless, in nocturnal rats, a majority of SCN neurons are stimulated by light, but in diurnal mice, a significant number of SCN neurons are suppressed by light. The findings indicate that the proportion of GABA to glutamate released from ipRGCs onto the SCN is significantly higher in nocturnal rats compared to diurnal rodents (Schoonderwoerd et al. 2022). The functional importance of these inhibitory projections in diurnal rodents is not yet understood, but they may provide a mechanistic explanation for the variations in photoentrainment observed between different species.

Additional inputs to the suprachiasmatic nucleus of nocturnal animals have been discovered by traditional retrograde tract-tracing techniques, as opposed to more contemporary methods, including viral vectors and genetically-encoded tracing (for a comprehensive review, see Saleeba et al. 2019). These include non-photic projections from the median raphe nucleus in the brainstem and projections from the IGL in the thalamus via the geniculo-hypothalamic tract that transmit both light-related and non-light-related information (Glass et al. 2010). Interneurons of the intergeniculate leaflet (IGL) release the neurotransmitter GABA as well as the neuropeptides neuropeptide Y (NPY), enkephalin, and neurotensin onto neurons located in the ventral suprachiasmatic nucleus to reset the circadian clock in response to light and, possibly, cues related to arousal (Morin and Blanchard 2001). In contrast, neurons in the median raphe nucleus release only serotonin onto the ventral SCN to reset the biological clock in response to behavioural arousal and feedback from locomotor activity (Deurveilher and Semba 2005). Conventional tracing techniques have verified that, similar to nocturnal rats, the suprachiasmatic nucleus of diurnal rodents receives neuropeptide Y (NPY) projections from the intergeniculate leaflet (IGL) and serotonin projections from the median raphe nucleus (Cohen et al. 2013). The diurnal tree shrew, which is closely related to primates, has a suprachiasmatic nucleus that receives monosynaptic projections from several non-photic brain regions, such as the locus coeruleus and periaqueductal grey (Ni et al. 2021).

In recent times, viral vectors have been employed to detect monosynaptic inputs into genetically-defined subsets of SCN neurons. VIP neurons receive inputs from various brain regions, such as the ventromedial hypothalamus (VMH), arcuate nucleus (ARC), medial preoptic area (MPOA), paraventricular thalamus (PVT), and paraventricular nucleus of the hypothalamus (PVN) (Todd et al. 2018). The extensive distribution of nerve fibres throughout the body is in line with several traditional studies that have shown that the suprachiasmatic nucleus receives most of its nerve supply from the retina, intergeniculate leaflet (IGL), and raphe nucleus. Additionally, around 40 different brain areas directly send nerve projections to the SCN (Moga and Moore 1997). Interestingly, SCN neurons that generate the neuropeptide cholecystokinin (CCK) receive direct inputs from the ARC, VMH, and PVN but do not receive inputs from the main structures that project to the SCN (retina, IGL, or raphe nuclei). Therefore, SCN CCK neurons do not react to light and do not play a role in the synchronisation of biological rhythms with the light-dark cycle (Hannibal 2010). The specific information being transmitted to CCK, VIP, and other SCN neurons through these "secondary" projections remains ambiguous. However, it is possible that this information encompasses details regarding the individual's level of arousal and motivation (Webb et al. 2013). Neurons in the SCN GRP region, but not SCN AVP, seem to receive projections from the IGL, dorsal raphe nucleus, and median raphe nucleus (Morin and Bengio 2005). However, these connections were identified by traditional tracing combined with immunohistochemistry. Scattered monosynaptic projections from dopaminergic neurons in the ventral tegmental area (VTA) target SCN neurons expressing dopamine receptor 1a. The functional association between the release of dopamine onto the SCN and photoentrainment, as well as the weight gain related to hedonic feeding, has been established (Grippo et al. 2020). Nevertheless, the sources and purposes of extra monosynaptic connections to these specific SCN subgroups remain undetermined.

2.2.2 The Pacemaker of the Clock: Suprachiasmatic Nuclei

In order for the circadian timing system to operate effectively, it is necessary for all the circadian clocks in the body to be synchronised with each other and with the 24-h day. This synchronization is achieved by the master circadian pacemaker, known as the suprachiasmatic nucleus (SCN). Similar to other cells, SCN neurons have the ability to produce independent circadian rhythms. However, SCN neurons possess distinct characteristics that are significant in multiple ways. Initially, they get direct visual information from the retina, enabling them to align with the diurnal rhythm (Morin and Bengio 2005). Furthermore, these organisms possess unique coupling systems organized based on their topography. These mechanisms enable them to maintain synchronization with each other even in the absence of light (Aton and Herzog 2005). Furthermore, they produce a distinct circadian pattern of neuronal firing rate, enabling them to coordinate the activity of other cells in the body through many direct and indirect channels (Gachon 2007). Therefore, the primary pacemaker in the suprachiasmatic nucleus (SCN) aligns itself with the light/dark cycle and then aligns other subordinate cellular oscillators. In addition, due to internal coupling, the suprachiasmatic nucleus (SCN) is able to provide a consistent output signal even when there is no light/dark cycle. This explains the "free-running" circadian rhythms of physiology and behaviour that continue under continuous conditions.

The suprachiasmatic nuclei (SCN) are a significant target of the ipRGC projections from the retina (Gooley et al. 2001; Hattar et al. 2006). The suprachiasmatic nucleus is a bilaterally symmetrical cluster of neurons situated in the anteroventral region of the hypothalamus, positioned next to the third ventricle and positioned somewhat superior to the optic chiasm (Klein et al. 1991). Each unilateral suprachiasmatic nucleus in the mouse consists of around 10,000 neurons divided into two anatomical subdivisions. The ventral section, known as the "core", is located next to the optic chiasm and gets input from the retina. The dorsal region, known as the "shell", partially surrounds the core and also receives input from it (Abrahamson and Moore 2001) (see Fig. 2.2). The core projects its contents heavily to the shell, while the shell only projects a minimal amount of information back to the core (Leak et al. 1999). The neuronal cell bodies of the SCN are rather tiny, measuring approximately 10 μm. They possess uncomplicated dendritic arbours and may be situated in close proximity to one another. Neurons in the core and shell subregions are differentiated based on their neurochemical composition. Each unilateral core consists of approximately 1100 neurons, with vasoactive intestinal polypeptide (VIP) in 10% of all SCN cells. Additionally, there are smaller populations of neurons carrying calretinin (900), neurotensin (NT; 700), and gastrin-releasing peptide (GRP; 500). The shell of the brain has approximately 2100 neurons that carry arginine vasopressin (AVP), which makes up about 20% of all SCN cells. Additionally, there are around 1100 neurons that contain angiotensin II and approximately 500 neurons that contain met-enkephalin. Neuropeptides are often found together with GABA in a majority of SCN neurons; however, a small number of cells may contain glutamate. A majority of connections between SCN neurons are GABAergic (Strecker et al. 1997). The intrinsic structure and neuropeptides of SCN exhibit significant variations among different species (Morin and Bengio 2005) and are more intricate than what the basic core/shell terminology implies.

The SCN is widely recognised as the primary circadian pacemaker in animals, supported by multiple lines of evidence (see Weaver 1998). Research conducted as early as the 1970s demonstrated that damage to the SCN results in the disruption of both behavioural and physiological circadian rhythms. The introduction of foetal SCN tissue into hosts with SCN lesions and no circadian rhythm restores the rhythmicity, and the length of the rhythm is determined by the transplanted tissue rather than the genetic makeup of the host (Ralph et al. 1990). The SCN also produces distinct circadian patterns in the frequency of spontaneous neuronal firing when physically separated, either in a living organism (Inouye and Kawamura 1979) or in a laboratory setting (Shibata et al. 1982). However, findings suggest that this may not be an exclusive characteristic (Guilding and Piggins 2007). The firing pattern in SCN neurons is partially controlled by the circadian regulation of potassium channels on the cell membrane (Kuhlman and McMahon 2006). Therefore, the

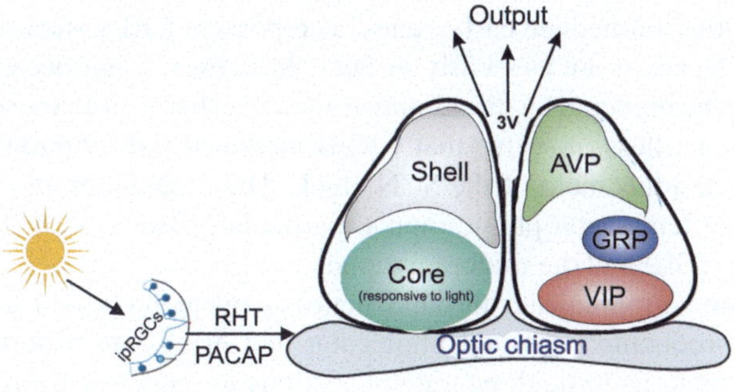

Fig. 2.2 Basic anatomical components of the mammalian SCN and important neuropeptides linked to it

suprachiasmatic nucleus (SCN) is typically both necessary and sufficient to express visible circadian rhythms in mice.

The input system of SCN plays a vital role in comprehending its operation. The retino-hypothalamic tract consists of axons from specialised photosensitive retinal ganglion cells (Berson 2003) that transmit signals to the suprachiasmatic nucleus (SCN). At the synaptic connections with SCN neurons, these axons release glutamate and pituitary adenylate cyclase-activating polypeptide (PACAP). The neuropeptide PACAP plays a regulatory role by augmenting the effects of glutamate (Michel et al. 2006). Glutamate stimulates the NMDA and AMPA receptors, causing membrane depolarization. This results in the entry of calcium ions, both directly through NMDA receptors and indirectly through voltage-sensitive channels (Meijer et al. 2012). Calcium triggers the activation of an intricate network of kinases (CamK, MAPK, PKA), resulting in the phosphorylation of CREB (Carlezon et al. 2005). Once activated, CREB stimulates transcription by attaching itself to calcium/cAMP response elements (CREs) found in genes. This process occurs in collaboration with additional proteins, such as CREB binding protein (CBP). Within the suprachiasmatic nucleus (SCN), phosphorylation of CREB at Ser133 and Ser142 takes place shortly after exposure to light, but only during the subjective night, when light pulses have the ability to alter the timing of the circadian clock (Gau et al. 2002). The circadian regulation of light input is accomplished, at least in part, by changes in sensitivity to glutamate at the receptor level (Pennartz et al. 2001). The induction of CREB targets in the SCN by light is specifically observed during the subjective night, as reported by CRE:βGal. The promoters of Per1 and Per2 genes include cAMP response elements (CREs) that specifically bind to the transcription factor CREB. These CREs are responsive to light stimulation in the suprachiasmatic nucleus (SCN). Mutant S142A mice, which lack the ability to undergo Ser142 phosphorylation, exhibit less attenuation of behavioural phase shifts (Gau et al. 2002). Additionally, the use of CREB-decoy (Tischkau et al. 2003), Per1 antisense, or Per2 antisense oligodeoxynucleotides (Wakamatsu et al. 2001) in the SCN also results in reduced attenuation of behavioural phase shifts. Since Per1 is activated within 5–15 min after exposure to light and plays a direct role in the central clock feedback mechanism, its activation could be primarily responsible for resetting the clock. Nevertheless, it is worth noting that additional CREB targets, including c-fos and other immediate early genes, also possess CREs and are activated by light in the SCN, albeit not as swiftly as Per1. Moreover, it has been observed that phase shifts can be hindered by the administration of c-fos + junB antisense (Wollnik et al. 1995). Regardless, it seems that CREB-mediated transcription is crucial for the light-induced adjustment of the SCN clock. The modulation of photic input is influenced by different non-photic inputs, particularly from the thalamic intergeniculate leaflet (IGL) and the midbrain raphé.

SCN neurons communicate through neurochemical processes to synchronise their activity, producing consistent molecular and firing rate patterns. They then transmit these patterns through neural connections to output pathways that control observable physiological and behavioural rhythms. The intercellular connection inside the suprachiasmatic nucleus (SCN) plays a crucial role in the functioning of the circadian clock (see Mohawk and Takahashi 2011). Multiple neurotransmitters, such as vasoactive intestinal peptide (VIP), gastrin-releasing peptide (GRP), and gamma amino butyric acid (GABA), play a crucial role in the connections among SCN neurons, as stated by Maywood et al. (2010). VIP is specifically present in a subgroup of neurons located in the central "core" area of the SCN. Neurons that carry VIP have extensive projections within the central region of the SCN, as well as to the surrounding "shell" area and even outside the limits of the SCN. Mice that do not have VIP, or the VPAC2 receptor for VIP, experience significant disturbances in their circadian rhythms (Aton and Herzog 2005; Colwell et al. 2003; Maywood et al. 2010). At the cellular level, the absence of VIP or VPAC2 signalling leads to a reduction in the amplitude of individual neuronal oscillators. This lack of signalling also results in a lack of coordination among the neurons. The results strongly indicate that a deficiency in coupling is responsible for the disrupted circadian rhythm in these strains of mice. The VIP/VPAC2 mechanism operates by increasing the levels of cAMP in the SCN, as demonstrated by Atkinson et al. (2007). Activation of intracellular signal transduction pathways is crucial for maintaining SCN neuronal synchronization (Brancaccio et al. 2013; O'Neill 2014). In Drosophila, the peptide dispersion hormone (PDF) is the equivalent of VIP in mammals. PDF and its receptor have a comparable function to mammalian VIP and the VPAC2 receptor in coordinating neurons in the circadian neural network of flies (Duvall and Taghert 2013). SCN neurons possess an independent circadian clock within their cells, which is regulated by many processes that contribute to the rhythmicity of ion channel activity, neuronal firing rate, and neurotransmitter release in the SCN. The regularity of potassium channel activity in the SCN is essential for generating strong circadian cycles in membrane potential and SCN firing rate (Hablitz et al. 2014; Itri et al. 2005; Kudo et al. 2011; Meredith et al. 2006). Disruption of the regular activity of potassium channels, specifically the large conductance "BK" channels or fast delayed rectifier current, results in a decrease in the amplitude of the beat. This is probably because individual SCN neurons have a decreased capacity to produce strong, self-sustained oscillations, and as a result, SCN neurons struggle to control the release of neurotransmitters. G protein-coupled inwardly rectifying potassium channels (GIRKs) exhibit rhythmic expression and are believed to significantly facilitate phase shifts caused by non-photic stimuli. This could be attributed to the crucial function of action potentials in the synaptic release of neurotransmitters in the suprachiasmatic nucleus (SCN), such as VIP, GRP, and GABA. The process of synaptic vesicle cycling is crucial for regulating circadian rhythms. Multiple proteins associated with synaptic vesicles are essential for maintaining circadian rhythmicity, as demonstrated by many studies (Deery and Jago 2009; Kapfhamer et al. 2002; Kim et al. 2009; Punia et al. 2012).

2.2.3 Outputs of the Clock: SCN Outputs

The suprachiasmatic nucleus (SCN) does not establish extensive synaptic connections across the brain (Abrahamson and Moore 2001). The primary monosynaptic connections originating from the SCN are directed towards a limited number of nuclei in the neighbouring hypothalamus and thalamus. These projections primarily originate from the periventricular region and target specific hypothalamic nuclei located near the third ventricle. These include the sub-paraventricular zone (SPZ), paraventricular nucleus (PVN), and dorsomedial hypothalamus (DMH). Additionally, there are projections that extend from the dorsal region to the rostral part of the paraventricular nucleus of the thalamus (PVT) (Kalsbeek et al. 1993; Watts et al. 1987). Using anterograde tracing techniques, researchers have found that SCN subpopulations in Vip-IRES-Cre, Avp-IRES2-Cre, and Prok2-EGFP mouse lines have distinct projection patterns (Sunkin et al. 2012). The DMH, SPZ, and PVT get a dense supply of axons from SCN neurons that express VIP, AVP, and PK2. Axons originating from VIP neurons exhibit prominent projections to various regions including the anterior hypothalamus (AHN), ventromedial preoptic nucleus (VMPO), medial preoptic area (MPOA), paraventricular nucleus (PVN), periventricular nucleus (PeVN), lateral habenula (LHb), lateral hypothalamus (LH), and ventromedial hypothalamus (VMH). Axons from VIP neurons have a limited distribution, projecting to specific regions such as the lateral septum, bed nucleus of the stria terminalis (BNST), ventrolateral preoptic nucleus (VLPO), median preoptic nucleus (MPN), anteroventral periventricular nucleus (AVPV), posterior hypothalamus (PH), and supraoptic nucleus (SON). The axons originating from AVP neurons have a significant projection to many brain regions, including the VMPO, organum vasculosum of the lamina terminalis (OVLT), AVPV, PVN, and arcuate nucleus (ARC). The efferent projections from other subpopulations of SCN neurons, such NMS+, GRP+, and CCK+, have not been thoroughly studied utilising contemporary genetic techniques. Interestingly, axons from the SCN NMS specifically target a distinct group of dopaminergic neurons in both the PeVN and PVN (Porcu et al. 2022).

The precise role of a majority of SCN projections to brain regions further down the neural pathway is not well understood. Nevertheless, a limited number of output circuits have been successfully traced from the suprachiasmatic nucleus (SCN) to regulate physiological and/or behavioural cycles. SCN VIP neurons have the ability to both stimulate gonadotropin-releasing hormone (GnRH)-producing neurons in the MPOA and inhibit gonadotropin inhibitory hormone (GnIH)-producing neurons in the DMH (Christian and Moenter 2008). SCN AVP neurons stimulate kisspeptin-producing neurons in the AVPV, activating GnRH neurons (Piet et al. 2015). When the daily timing signal from the suprachiasmatic nucleus (SCN) aligns with elevated oestrogen levels, the gonadotropin-releasing hormone (GnRH) neurons stimulate the release of luteinizing hormone from the anterior pituitary. This hormone spike then initiates ovulation, which takes place in the mid-afternoon every 4–5 days in nocturnal rodents (Willison et al. 2013). The suprachiasmatic nucleus controls the oscillations of body temperature by sending direct AVPergic projections to the median preoptic nucleus (MnPO) and indirect projections to the arcuate nucleus (ARC) (Guzmán-Ruiz et al. 2015). During the period of activity (which occurs at night in nocturnal rats), the suprachiasmatic nucleus stimulates arcuate nucleus (ARC) neurons to secrete α-melanocyte stimulating hormone (α-MSH) onto the median preoptic nucleus (MnPO) to maintain a high body temperature. During the period of inactivity, the suprachiasmatic nucleus (SCN) suppresses the release of alpha-melanocyte-stimulating hormone (α-MSH) from the arcuate nucleus (ARC) while also directly releasing arginine vasopressin (AVP) onto the median preoptic nucleus (MnPO) to decrease body temperature. Due to the ARC's ability to encode metabolic information, thermoregulatory cycles can be flexibly adapted according to an animal's metabolic requirements (Buijs et al. 2019). The mechanism by which the ARC integrates circadian and metabolic information remains unknown.

The suprachiasmatic nucleus (SCN) sends projections to the paraventricular nucleus (PVN) both directly and indirectly via the subparaventricular zone (SPZ). This communication is responsible for regulating several rhythms in the body, such as the release of melatonin, glucocorticoids, and heart rate (Perreau-Lenz et al. 2004). The multi-synaptic pathway for melatonin release begins with neurons in the suprachiasmatic nucleus (SCN) that send projections to pre-autonomic paraventricular nucleus (PVN) neurons. These PVN neurons then form synapses with neurons in the intermediolateral column of the spinal cord (IML). The IML neurons send projections to noradrenergic neurons located in the superior cervical ganglion, which in turn activate the pineal gland to secrete melatonin (Perreau-Lenz et al. 2004). The PVN receives a consistent degree of excitatory input from the sparsely distributed glutamatergic projections originating from the SCN. Nevertheless, the strength of this signal is overshadowed by the GABAergic SCN projections that cyclically suppress PVN neurons (and, therefore, melatonin secretion) during daylight hours in all vertebrates (Perreau-Lenz et al. 2004). Nevertheless, in animals that are active during the night and those that are active during the day, the action of PVN neurons that produce corticotropin-releasing hormone (CRH) stimulates the secretion of adrenocorticotropic hormone (ACTH) from the pituitary gland. This, in turn, triggers the adrenal glands to release a maximum amount of glucocorticoids right before awakening, either at dusk or dawn, depending on the animal's activity pattern.

The adrenal glands also receive indirect projections from pre-autonomic PVN neurons that regulate their responsiveness to ACTH (Jasper and Engeland 1994). In mice active throughout the night, a group of neurons called SCN AVP neurons suppress another group called PVN neurons. This suppression can occur either directly or indirectly through the action of GABAergic interneurons in the SPZ (Kalsbeek et al. 1993). It is hypothesised that in animals that are active throughout the day, the SCN AVP neurons connect to excitatory glutamatergic SPZ interneurons. SCN VIP neurons exert an inhibitory effect on the PVN, leading to the suppression of glucocorticoid production. This is most likely achieved through the paracrine release of VIP and/or the synaptic release of GABA (Mazuski et al. 2020). Currently, the specific stimulatory signal from the SCN remains unknown. However, it is possible that this signal is not essential for promoting the release of glucocorticoids. This is because the inherent rhythms of PVN neurons may be enough to override the weak inhibitory input from the SCN (Jones et al. 2019). The inhibitory projections from the suprachiasmatic nucleus (SCN) to the paraventricular nucleus (PVN) also have a role in regulating diurnal patterns in autonomic function, particularly heart rate.

The SCN AVP and VIP neurons send both direct and indirect connections to the PVN, producing AVP and GABA to suppress the activity of pre-autonomic PVN neurons. The PVN neurons have direct and indirect connections, via the rostral ventrolateral medulla, with neurons in the IML. These IML neurons then project to noradrenergic neurons in the stellate ganglion, which in turn innervate the heart. Additional autonomic rhythms, such as hepatic glucose production, blood pressure, and adipocyte thermogenesis, are probably regulated by the same or comparable circuitry that controls the suprachiasmatic nucleus (SCN) output.

Sleep, a prominent circadian behaviour, is controlled by a "two-process model" where a circadian want to be awake counteracts a homeostatic requirement for sleep (Borbély 1982). Although the specific neural processes responsible for sleep homeostasis remain unclear, there has been some progress in understanding the circadian circuitry that controls wakefulness. The SCN sends projections to the DMH directly and indirectly via the ventral SPZ, not the dorsal SPZ (Vujovic et al. 2015). The GABAergic projections from the dorsomedial hypothalamus (DMH) suppress the activity of neurons in the ventrolateral preoptic nucleus (VLPO) that promote sleep. At the same time, the glutamatergic projections from the DMH boost the activity of hypocretin neurons in the lateral hypothalamus (LH), which promote wakefulness. The activation of the ascending reticular activating system, which includes a cholinergic pathway from the brainstem and a monoaminergic pathway from the brainstem and caudal hypothalamus, finally stimulates the cortex to enhance and maintain awake (Scammell et al. 2017). The multi-stage circadian output route provides adaptability in sleep scheduling and may elucidate the variations in sleep timing observed in nocturnal and diurnal species.

Damage to the DMH or the ventral (but not dorsal) SPZ significantly impairs circadian rhythms in locomotor activity (Tahara et al. 2010). SCN VIP neurons exhibit functional connectivity with DMH neurons, as the optogenetic stimulation of VIP neurons results in the suppression of firing activity in a specific group of SPZ neurons that project to the DMH. As a result, the duration and intensity of locomotor activity patterns are reduced when SCN VIP neurons are genetically removed (Mazuski et al. 2020). The DMH, like other hypothalamic nuclei, consists of a diverse group of neurons. However, we have yet to fully comprehend which specific subpopulation(s) of DMH neurons are responsible for regulating locomotor activity cycles. The locomotor activity cycles remain unchanged when chemogenetically stimulating or genetically ablating GABAergic DMH neurons, but they are completely eliminated when leptin receptor-expressing DMH neurons are genetically silenced (Li and Liu 2016). Nevertheless, the precise impact of these distinct subtypes of DMH neurons on locomotor activity rhythms remains unclear (Zhou et al. 2022). The regulation of locomotor activity is also controlled by the projections from the suprachiasmatic nucleus (SCN) to the arcuate nucleus (ARC). Severing the connection between the SCN and the ARC through surgery, while keeping both nuclei intact and functional, eliminates the regular patterns of locomotor activity in complete darkness (Buijs et al. 2019). The precise cellular constituents of the SCN-ARC circuit remain uncertain. However, it is probable that they encompass neuropeptide Y (NPY) and agouti-related peptide (AgRP)-co-expressing neurons, as well as kisspeptin-expressing neurons (Coppari et al. 2005).

The DMH (dorsomedial hypothalamus) and ARC (arcuate nucleus) are also involved in regulating the timing of food consumption according to the body's internal clock. The DMH was previously suggested as the location of the "food entrainable oscillator", which is responsible for the continuation of food anticipatory activity following damage to the SCN. Recent investigations have determined that the DMH specifically regulates meal anticipatory activity, even after total DMH ablation (Acosta-Galvan et al. 2011). Nevertheless, the DMH plays a crucial role in the route that connects the SCN to feeding behaviour. Lesions in the DMH have the effect of disrupting or totally eliminating circadian feeding rhythms (Bernardis 1973). The precise cellular subtypes of the DMH that contribute to the regulation of eating rhythmicity remain incompletely elucidated. Suppressing tropomyosin receptor kinase B (TrkB)-expressing DMH neurons via chemogenetics leads to an elevation in food consumption, specifically during the light (inactive) phase, but not during the dark (active) phase of a light:dark cycle (Liao and Rust 2021). Furthermore, genetically suppressing DMH neurons that express the leptin receptor leads to disturbances in the daily patterns of food consumption (Faber et al. 2021). The function of these neurons is to suppress the activity of orexigenic NPY/AgRP co-expressing ARC neurons, which play a crucial role in controlling eating behaviour (Rau et al. 2021). The circadian patterns of food intake are eliminated in mice that have undergone targeted chemical removal of either NPY/AgRP neurons or leptin receptor-expressing ARC neurons. These ARC neurons consist of both NPY/AgRP neurons and α-MSH-expressing neurons, which are a subset of anorexigenic pro-opiomelanocortin (POMC)/cocaine and-amphetamine-regulated transcript (CART) co-expressing ARC neurons (Wiater et al. 2011). While it is known that the suprachiasmatic nucleus (SCN) has a strong connection with the arcuate nucleus (ARC), it remains uncertain if the circadian clock directly affects the POMC/CART or NPY/AgRP neurons. This uncertainty arises because these neurons often collaborate to either stimulate or inhibit feeding behaviour. Suppressing the genetic activity of kisspeptin-expressing ARC neurons also disturbs the patterns of feeding, most likely by interfering with their usual suppression of NPY/AgRP and stimulation of POMC/CART signalling.

Although it is evident or highly probable that more intricate behaviours are rhythmic in nature, the specific neural connections between the suprachiasmatic nucleus (SCN) and the brain regions responsible for controlling these behaviours remain largely uncharted. Nevertheless, a complex neural pathway responsible for a specific behavioural pattern, aggression, has been identified, starting from the suprachiasmatic nucleus (SCN), passing through the subparaventricular zone (SPZ), and ending in the ventromedial hypothalamus (VMH) (Todd et al. 2018). Within this circuit, SCN VIP neurons provide projections to GABAergic neurons located in the dorsal SPZ, which in turn innervate neurons in the ventrolateral VMH. The occurrence of aggressive behaviour in mice typically reaches its highest point at the start of the dark (active) phase. However, when the vesicular GABA transporter (VGAT) is specifically removed in SPZ neurons, the timing of this rhythm moves to the beginning of the light (inactive) phase. Thus, chemogenetically suppressing GABAergic SPZ neurons leads to heightened aggression during the early morning hours but does not produce any impact throughout the evening. Ultimately, other indirect connections have been discovered between the SCN and brain areas responsible for controlling wakefulness, drive, and emotional state. As an illustration, the suprachiasmatic nucleus sends signals to the ventral tegmental area (VTA) through several synapses, including the dorsomedial hypothalamus (DMH), lateral habenula (LHb), and medial preoptic nucleus (MPN). It also communicates with the locus coeruleus through the medial preoptic area (MPOA) and DMH, the dorsal raphe nucleus through the DMH, and the nucleus accumbens through the VTA and paraventricular nucleus of the thalamus (PVT) (Conway-Campbell et al. 2010). The precise cellular components participating in these output circuits of the SCN, as well as the mechanisms by which they use timing information from the SCN to regulate behaviour in a rhythmic manner, remain unknown.

2.3 The Molecular Mechanisms Underlying Circadian Oscillation

The intracellular oscillation mechanism involves a feedback loop between transcription and translation processes. Activating transcription factors generally enhance the expression of genes that produce products that then suppress their own production, resulting in a feedback loop with a cycle length of around 24 h. The mRNAs and proteins that exert the most dynamic control over the clock undergo synthesis and degradation within each circadian cycle. Multiple, parallel, and interlocking feedback loops exist that collectively strengthen rhythmicity, enhance cycle length precision, and amplify rhythm amplitude. The chemicals and mechanisms governing circadian rhythms exhibit a high degree of similarity between flies and mammals (Weaver and Emery 2013).

The transcriptional activator complex in mammals comprises heterodimers that contain basic helix-loop-helix (bHLH) and Per-ARNT-SIM (PAS) domains. The bHLH-PAS proteins CLOCK and BMAL1 form a heterodimeric complex that plays a crucial role in the activity of the circadian oscillator (BMAL1 is also known as MOP3 and ARNTL). Within a specific group of neurons in the SCN, a protein called neuronal PAS domain-containing 2 (NPAS2) is present. NPAS2 is a bHLH-PAS protein that shares a similar sequence to CLOCK. It has the ability to sustain circadian rhythms even when CLOCK is not present, most likely by interacting with and stabilizing another protein called BMAL1 (Debruyne 2008). Mice that do not have the BMAL1 gene or neither of BMAL1's partner genes (CLOCK and NPAS2) do not have circadian rhythms (Bunger et al. 2000; Debruyne 2008). NPAS2 is expressed at low levels in peripheral tissues, so it cannot compensate for the lack of CLOCK.

Multiple studies have discovered numerous direct transcriptional targets of CLOCK: BMAL1 complexes (Hatanaka et al. 2010; Rey et al. 2011). The CLOCK:BMAL1 complex regulates the expression of three Period genes (Per1,

Per2, and Per3) and two Cryptochrome genes (Cry1 and Cry2). The protein derived from the Per3 gene, known as PER3, appears to have a limited impact on circadian rhythm (Shearman et al. 2000). On the other hand, PER1 and PER2 play a crucial role. There is a degree of overlap in function between PER1 and PER2, as well as between CRY1 and CRY2. Simultaneous removal of either Per1 and Per2, or Cry1 and Cry2, results in the absence of circadian rhythmicity (Bae et al. 2001; Zheng et al. 2001). The PER and CRY proteins produce essential complexes crucial for negative feedback. The significant oscillatory manifestation of PER1 and PER2 is the crucial factor that determines the speed at which PER:CRY complexes are formed (Lee et al. 2010). The continuous expression of CRY does not interfere with the functioning of the oscillator; however, the continuous expression of PER does. However, the control of both PERs and CRYs' stability directly affects the duration between the activation of transcription and the subsequent negative feedback, hence determining the length of the circadian oscillation.

Other than the genes implicated in the fundamental feedback loop, numerous other genes exhibit rhythmic expression in a manner contingent upon the circadian clock. Approximately 10% of the active genes in the analysed tissues show rhythmic expression. These genes exhibiting rhythmic expression are often important for controlling essential processes and are considered critical steps in these processes (Akhtar et al. 2002; Panda et al. 2003).

Certain clock-regulated genes are directly regulated by CLOCK/NPAS2: BMAL1 complexes, alongside the PER and CRY genes. However, these genes do not play a role in the feedback loop or transcriptional networks. The presence of arginine vasopressin (AVP) in the suprachiasmatic nucleus (SCN) serves as a prominent illustration of an output gene that is directly controlled at the transcriptional level by the CLOCK:BMAL1 complex (Jin et al. 1999). This neuropeptide is released in a rhythmic manner at specific locations in the neurons and into the cerebrospinal fluid. It has an impact on brain function, but it does not participate in a molecular feedback loop that affects its own regulation. Additional genes that are controlled by CLOCK/NPAS2:BMAL1 heterodimers also function as transcription factors. They have the ability to function as either transcriptional activators or transcriptional repressors. Secondary, interlocking feedback loops are formed by members of four transcription factor families: nuclear hormone receptors, PAR b-Zip, DEC1/DEC2, and Chrono. The interlocking loops serve to stabilize and strengthen the major CLOCK/NPAS2:BMAL1:PER1/PER2:CRY1/CRY2 feedback mechanism. A notable characteristic of these transcription factor networks is that the activator complex frequently induces the expression of genes within the same gene family that possess contrasting activity.

The rivalry and opposition between activators and repressors for promoter elements is a crucial and enduring design feature in the circadian clock. The transcription factors in the interlocking feedback loops influence three key response elements involved in rhythmic transcription: E- and E-like boxes (activated by bHLH-PAS proteins and CHRONO), D-box elements (activated by the PAR bZIP family), and RORE elements (activated by orphan nuclear hormone receptors) (Ueda et al. 2005). The sequential activation of these transcriptional factor cascades ultimately results in the regulation of the expression of secondary genes that are influenced by their activity, albeit with a slight delay. Some of these genes that are regulated by the body's internal clock are, once again, transcription factors. Therefore, similar to the way ripples expand from a stone tossed into a pond, CLOCK/NPAS2:BMAL1 heterodimers trigger progressively broader patterns of gene expression, resulting in a range of phases and profiles of gene activity. Post-transcriptional processes offer additional means of control. It has been shown that four protein families create interconnected feedback loops that function simultaneously with the CLOCK:BMAL1: PER:CRY core feedback loop. These modifiers are usually not necessary for circadian oscillation, but they can play significant roles in altering the amplitude and phase of gene expression rhythm. Multiple nuclear orphan receptors, namely Reverb-α, ROR-α, and ROR-γ-t, are instances of CLOCK/NPAS2:BMAL1 targets that control the activity of other genes. One of the genes that are regulated by the nuclear orphan family members through the activation of nuclear orphan receptor response elements is Bmal1. The RORs stimulate, while REV-ERB-α inhibits, the expression of the Bmal1 gene, creating a mutually dependent feedback mechanism in the circadian clock (Preitner et al. 2002; Sato et al. 2004; Ueda et al. 2005). Excessive expression of REVERB-α inhibits the expression of Bmal1.

The PAR bZIP transcription factors, which belong to a family of proteins rich in proline and acidic amino acids and contain a basic leucine zipper, are involved in an interconnected feedback loop (Gachon 2007). The PAR bZIP transcription factors consist of the activators DBP (D-box binding protein), TEF (thyrotroph embryonic factor), HLF (hepatic leukaemia factor), and the repressor E4BP4 (E4 promoter-binding protein 4). The expression of these genes is enhanced by CLOCK:BMAL1 complexes, and they exhibit rhythmic expression (Gachon 2007; Ripperger and Merrow 2011). Their protein products exert their effects by interacting with D-box regions to control gene expression (Ueda et al. 2005). Per1 and Per2 are genes that are regulated by rhythmically expressed transcription factors. This regulation creates a potential interlocked feedback loop (Mitsui et al. 2001; Yamaguchi et al. 2000). Additionally, DBP and E4BP4 play a crucial role in determining the period of these genes. In the liver, this gene family regulates the expression of genes related to the synthesis of bile acids, such as cholesterol 7-α hydrolase (Cyp7a), as well as genes involved in the metabolism of foreign substances (Gachon 2007). The circadian control of PAR bZIP genes plays a crucial role in governing the daily patterns of drug reactions.

DEC1 and DEC2, two bHLH-PAS proteins, are clock-controlled genes that impact the oscillatory process. The DECs disrupt the activity of CLOCK: BMAL1 by competing for E-box elements in genes that respond to them. Additionally, they can interact with CLOCK:BMAL1 protein complexes to hinder their action. The lack of DEC1 and DEC2 has a limited impact on both molecular and behavioural rhythmicity (Rossner et al. 2008). The gene formerly referred to as Gene model 129 (Gm129) plays a crucial role in controlling circadian rhythms. Gm129 has been given the new name "Chrono" due to its function as a circadian modulator (Anafi et al. 2014; Goriki et al. 2014). Chrono exhibits cyclical expression, binding to E-box elements in a rhythmic manner and suppressing the expression of genes triggered by CLOCK:BMAL1 (Anafi et al. 2014; Annayev et al. 2014; Goriki et al. 2014). CHRONO interacts with both BMAL1 and the glucocorticoid receptor. These papers present conflicting views of whether CHRONO suppressed CLOCK:BMAL1-induced gene expression by interfering with the interaction with the coactivator protein CBP or modifying the interaction with histone deacetylase 1 (HDAC1). Studies have shown that mice lacking CHRONO, particularly in vasopressin neurons, and mice with Chrono knockout have a slight but statistically significant increase in circadian period under constant conditions. However, another study mentioned that their experiment did not have enough power to detect small differences in the free-running period (Annayev et al. 2014). These findings collectively demonstrate that Chrono is a gene product regulated by a clock mechanism, which contributes to the formation of an additional interconnected feedback loop.

The establishment of a molecular feedback loop, which entails the processes of transcription and translation and subsequent negative feedback on transcription, can be readily achieved within a short span of a few hours. As an illustration, various stimuli trigger the production of the c-fos gene in the brain. The highest levels of gene transcripts are observed 20–30 min after the stimulation, while protein levels reach their peak around 1–2 h. Furthermore, the biological clock that controls somites during development exhibits a transcriptional-translational feedback loop structure comparable to the circadian clock. However, this cycle has a period duration of 2 h (Hirata et al. 2002). To attain the approximately 24-h duration of circadian rhythms, it is necessary to meticulously and extensively control the chemical processes occurring in rhythmic cells. As mentioned before, the negative feedback effectors for circadian rhythms consist of complexes formed by PER and CRY proteins.

Phosphorylation, ubiquitination, and proteasomal degradation play crucial roles in controlling the duration of their existence, and altering these mechanisms influences the pace of the circadian clock. For insights into the influence of these posttranslational mechanisms on circadian rhythm control, refer to studies by Gallego and Virshup and Reischl and Kramer (Gallego and Virshup 2007; Reischl and Kramer 2011). The PER proteins undergo phosphorylation, ubiquitination, and proteasomal degradation by several protein kinases. The kinases responsible for the posttranslational regulation of PERs are casein kinase 1 (CK1) δ and CK1 ε. The disruption of CK1 δ has a noticeable but meaningful impact on the period of circadian rhythm, resulting in an increase of approximately 1–2 h. This effect has been observed in both cellular and tissue assay systems (Etchegaray et al. 2006). A mutation of CK1 δ in humans is a known cause of familial advanced sleep phase syndrome (Xu et al. 2005). CK1 δ has a greater significance in regulating the period than CK1 ε. Drugs that affect both isoforms have a stronger impact compared to the genetic disruption of either isoform alone or selective suppression of CK1 ε (Etchegaray et al. 2009; Meng et al. 2008; Walton et al. 2009). Substances that lack selectivity for the CK1 isoform significantly prolong the duration of the circadian cycle, indicating that there is functional similarity between CK1 δ and CK1 ε (and maybe other kinases).

Furthermore, the results of small chemical screens indicate that simultaneous inhibition of many kinases leads to a much-enhanced response (Hirota et al. 2010). The tau mutation, a spontaneous mutation of CK1 ε, has been discovered in Syrian hamsters. This mutation results in a reduced period (Lowrey et al. 2000; Ralph et al. 1990). The introduction of this mutation into CK1 ε in mice results in a comparable semi-dominant, short-period phenotype (Meng et al. 2008). The tau mutant CK1 ε exhibits enhanced activity in phosphorylating PERs and facilitating their breakdown (Gallego et al. 2006). The period length regulation depends on the presence of PER proteins, which are influenced by the casein kinases (Maywood et al. 2010). The process of phosphorylating the PER proteins by casein kinases is counteracted by protein phosphatases, particularly protein phosphatase 1 (Schmutz et al. 2011) and protein phosphatase 5 (Partch et al. 2014). Maintaining this equilibrium is crucial for determining the speed of the biological clock. Phosphorylation occurs before and allows for ubiquitination by E3 ubiquitin ligases, which facilitates proteasomal breakdown. Unsurprisingly, the activity of the circadian clock is influenced by many F-box proteins and the ubiquitin-proteasome-related process. The F-box proteins β-TrCP1 and B-TrCP2 function as ubiquitin ligases, which means they facilitate the attachment of ubiquitin to substrates such as PER1 and PER2, leading to their destruction through the proteasome (Ohsaki et al. 2008). The breakdown of PER can be hindered by altering B-TrCP or by blocking proteasomal degradation (e.g., with MG132), resulting in an increase in the duration of the circadian period. CRY proteins are subjected to degradation by two F-box proteins, namely FBXL3 and FBXL21. Loss-of-function mutations in Fbxl3 result in longer circadian periods (Busino et al. 2007). These proteins engage in competitive and antagonistic interactions with each other (Hirano et al. 2013; Yoo et al. 2013). Although CRY1 and CRY2 both block CLOCK:BMAL1-mediated transcription, animals with disrupted versions of both genes exhibit contrasting effects on the length of the period under constant darkness (Vitaterna et al. 1999).

The relative activity levels of the two CRY proteins, with CRY1 being more active than CRY2 as a transcriptional repressor, and the relative levels of the two F-box proteins, are crucial factors in determining the speed of the biological clock (Anand et al. 2013). A high-throughput screening of small molecules was conducted to identify compounds that can modify the duration of circadian oscillations in a laboratory setting. One of the compounds discovered in this screen is KL001. It was shown that KL001 interacts with CRY proteins and inhibits their degradation by the proteasome. This interaction results in an elongation of the period length in cellular experiments (Hirano et al. 2013). Although the main topic of this discussion has been the stability of the PER:CRY complex, which is the most dynamic and time-regulated component of the circadian oscillator, it is important to note that these kinases also target other proteins involved in the circadian clock. Secondary phosphorylation occurs inside complexes consisting of kinase:PER:CRY, and phosphorylation can also occur directly when kinases are produced alongside other circadian proteins in laboratory settings. The CLOCK protein experiences periodic alterations in its phosphorylation status (Lee et al. 2010). These variations have an impact on the protein's functionality and durability. Lithium, a pharmacological agent employed for managing bipolar disorder in people, has been widely recognized for its ability to extend the duration of circadian rhythms in cells, tissues, and entire organisms.

Glycogen synthase kinase 3 (GSK3) has been widely believed to be the specific biological target of lithium. Lithium has the ability to inhibit GSK3 (Can et al. 2014). Suppressing GSK-3 disrupts the stability of REV-ERB-α, thereby impacting the production of BMAL1 and other elements of the molecular circadian rhythm (Iitaka et al. 2005). GSK3-β has been suggested as a kinase that has a role in phosphorylating CLOCK (Spengler et al. 2012) and BMAL1 (Sahar et al. 2010). Lithium suppresses GSK-3-β and prolongs the circadian period, but more specific inhibitors of GSK-3-β reduce the circadian period (Hirota et al. 2010). The activation of GSK-3 has been found to lengthen the circadian period. The mechanism by which both raising and reducing GSK-3 activity might lengthen the period is not clearly understood, which implies that there may be other targets for lithium that are crucial to the biological clock (Paul et al. 2012). Lithium also hinders the activity of myo-inositol monophosphatase (IMPase). Mice with a specific genetic mutation that causes a loss of enzymatic activity in this gene experience a slight increase in the length of their circadian period (Ohnishi et al. 2014). This finding, indicates that targeting IMPase alone is not enough to fully explain the effects of lithium in lengthening the circadian period.

The Drosophila timeless gene, known as dTIM, plays a crucial part in the molecular clock of flies. It functions alongside Drosophila dPER in the repressor complex, similar to how CRY1 and CRY2 function in mammals (Weaver and Emery 2013). Before the discovery of the involvement of mammalian CRYs, it was believed that mammalian TIM (mTIM), a homolog found in mammals, played a crucial role as a binding partner for PERs and as a component of the circadian clockwork in mammals. The precise function of mTIM in regulating the circadian rhythms of mammals is still uncertain (Gotter et al. 2007). mTIM exhibits a higher degree of genetic similarity to Drosophila timeout compared to dTIM. Although mTIM is present in complexes that include CRY proteins and can hinder the transcriptional activation mediated by CLOCK:BMAL1 in mammalian cells, it is unable to do so when produced in S2 cells. This suggests that its interactions with the activator complex may be indirect. Mice who do not have mTIM die before reaching midgestation, which prevents the possibility of doing behavioural study. TIM has additional functions that are not related to the circadian rhythm, but rather involve the structure of chromosomes (Gotter et al. 2007). Engelen et al. (2013) found that mTIM influences the length of the circadian cycle and has a role in the impact of DNA damage on the circadian phase. In a previous study (Barnes et al. 2003) it was suggested that mTIM plays a special role in controlling the rhythmicity of the SCN. mTIM potentially participates in the regulation of circadian rhythms by modulating the availability of CRY. The latest evaluations of the mammalian circadian system do not address mTIM, indicating that its function remains ambiguous to a majority of researchers.

2.3.1 Transcription Factors and Cofactors

Several accessory proteins, cofactors, and enzymes that modify histones play a role in the molecular oscillator that controls circadian rhythms. Several proteins engage in interactions with CLOCK and/or BMAL1, hence exerting an impact on their function as transcriptional activators. The complexes that contain PER:CRY molecules exert transcriptional repression by binding corepressors and/or inhibiting the activity of CLOCK:BMAL1-recruited transcriptional co-activators. This section will provide an account of certain interactions and the proteins involved in them (Chong et al. 2018; Partch et al. 2014). An effective method for discovering more clock-regulatory elements is isolating proteins that are found within complexes with other circadian clock proteins. NONO, which stands for Non-POU domain containing octamer-binding protein, was identified as a constituent of the PER:CRY repressive complex (Brown et al. 2005). Further research demonstrated that the DBHS gene family, consisting of NONO, SFPQ, and PSPC1, have overlapping functions in regulating circadian rhythms through E-box elements found in the promoters of circadian genes (Guillaumond et al. 2012; Kowalska et al. 2013). Additionally, NONO is responsible for linking the circadian clock with the cell cycle. CAVIN-3 is a protein that interacts with PER2 and affects the length of the circadian cycle by regulating the interactions between PER and CRY proteins (Schneider et al. 2012). BMAL1-containing complexes include Receptor for Activated C Kinase-1 (RACK1) and protein kinase C-α (PKC-α) during the negative feedback phase of the cycle (Robles et al. 2017). These components inhibit the transcriptional activity of CLOCK:BMAL1, maybe via phosphorylating BMAL1. Downregulating the expression of RACK1 or PKC-α resulted in a decrease in the length of the circadian period (Robles et al. 2017).

Preliminary evidence suggests that the protein modification enzyme LSD1 is a significant focus of PKC-α. Hence, the PKC signalling pathway plays a role in the circadian feedback loop by influencing both circadian clock proteins and chromatin. BMAL1-containing complexes additionally include TRAP150, also known as thyroid hormone receptor-associated protein-150 (Lande-Diner et al. 2015). TRAP150 exhibits rhythmic expression and facilitates the binding of CLOCK:BMAL1 to the promoters of target genes, hence boosting transcriptional activation. The absence of TRAP150 reduces the effectiveness of transcriptional activation and results in rhythms with low amplitude and extended periods. The isolation of CLOCK-containing protein complexes resulted in the discovery of "CLOCK-Interacting Protein, Circadian" (CIPC) as an additional negative feedback regulator of the circadian clock (Zhao et al. 2010). CIPC exhibits rhythmic expression and hinders the action of CLOCK:BMAL1. The exhaustion of CIPC reduces the duration of the circadian period. Heat-shock factor 1 (HSF1) and Upstream Stimulating Factor 1 (USF1) have been recognized as circadian transcription factors (Reinke and Asher 2019). The expression of USF1 varies depending on the strain and has a role in suppressing the long-period phenotype in ClockΔ19/+. The melatonin production route can inhibit the phenotype of ClockΔ19/+ by affecting the synchronization of SCN oscillators rather than through a transcriptional mechanism (Shimomura et al. 2010).

2.4 Chromatin Remodelling and the Epigenetics of Circadian System

Nucleosomes are structures in chromatin where DNA is coiled around complexes of histone proteins. The accessibility of DNA to transcription factors is significantly influenced by posttranslational changes in histones. The "histone code" exerts epigenetic regulation by influencing gene expression. This regulation occurs through the provision of additional information by histones, which goes beyond the elements included in the DNA coding sequence.

Various posttranslational modifications of histones, such as phosphorylation, acetylation, methylation, ADP-ribosylation, and ubiquitination, have the ability to impact gene expression. (For a comprehensive overview, see Marmorstein and Trievel 2009). These alterations that occur at specific sites are facilitated by particular enzymes, and they also happen in a collaborative manner (for example, phosphorylation at one residue can enhance the likelihood of acetylation at another residue). Various histone-modifying enzymes have been demonstrated to impact circadian rhythms (Masri and Sassone-Corsi 2013; Ripperger and Merrow 2011).

Light exposure, which resets the circadian rhythms, leads to an overall increase in the phosphorylation of histone H3 at the Serine residue 10 in the mouse SCN (Crosio et al. 2000). Nevertheless, the regulation of histone acetylation is controlled by the contrasting actions of histone transferases (HATs) and histone deacetylases (HDACs). The presence of HAT activity in CLOCK has been documented (Doi et al. 2006). Acetylation of histone H3, namely in the promoter-associated histones of rhythmically regulated genes, is observed, while histone H4 acetylation does not occur in this context (Etchegaray et al. 2009). The enzyme SIRT1, which relies on NAD(+) for its action, acts as a histone deacetylase, opposing the histone acetyltransferase (HAT) activity of CLOCK (Asher et al. 2010; Nakahata et al. 2009). This method allows cellular redox and metabolic status to impact circadian rhythmicity. CLOCK also controls transcriptional activity by acetylating proteins, such as BMAL1 (Hirayama et al. 2007) and the glucocorticoid receptor (Nader et al. 2010). The histone acetyltransferase CBP/p300 is found in complexes that contain CLOCK in liver nuclear extracts. The presence of this complex at Per gene promoters follows a cyclical pattern (Etchegaray et al. 2009). The transcriptional activity of CLOCK:BMAL1 is augmented by CBP/p300 and diminished by CRY proteins. CRYs counteract the function of CLOCK-containing complexes by inhibiting HAT activity and modifying chromatin structure, hence suppressing transcriptional activation. This results in a negative feedback loop. A study by Hosoda et al. (2009), found that the activity of p300 can have either an enhancing or inhibitory effect on CLOCK. The transcriptional activity mediated by BMAL1 varies depending on the cell line, and this variation is associated with the levels of the corepressor histone deacetylase 3 (HDAC3) and the coactivator pCAF in those cell lines.

The process of transcriptional suppression by the PER complex includes the formation of a complex consisting of PER1, SFP, and SIN3A. This complex then recruits HDAC1, which results in the removal of acetyl groups from the H3K9 protein at CLOCK:BMAL1 target sites, specifically the Per1 promoter (Duong et al.

2011). The nuclear orphan receptor REV-ERB-α (NR1D1) is no longer considered an "orphan" receptor because its ligand, heme, has been found (Yin et al. 2007). The Rev-erb-α gene has rhythmic expression in several organs, where it directly suppresses the expression of Bmal1 (Kornmann et al. 2007). REV-ERB-α additionally controls several metabolic genes. REV-ERB-α inhibits transcription by enlisting a Nuclear CoRepressor-HDAC3 complex, resulting in periodic histone deacetylation and extensive impacts on gene expression (Alenghat et al. 2008; Feng and Lazar 2012). Disruption of both Rev erb-α and Rev-erb-β results in the disturbance of rhythmic clock gene expression and circadian behavioural rhythms. This disruption also causes abnormal lipid metabolism and the development of hepatic steatosis (Bugge et al. 2012; Cho et al. 2020). Therefore, the rhythmic manifestation of REV-ERB-α is crucial in connecting circadian rhythms with metabolism. This occurs through rhythmic epigenetic regulation, facilitated by alterations in histone proteins (Masri and Sassone-Corsi 2013).

Additional metabolic factors that affect the functioning of the circadian clock interact with changes in histone modifications (as discussed in (Jordan and Lamia 2013). Nicotinamide phosphoribosyltransferase (NAMPT) is the enzyme that restricts the rate of NAD+ production in mammals. The expression of the Nampt gene, the activity of NAMPT, and the levels of NAD+ exhibit a rhythmic pattern (Nakahata et al. 2009; Ramsey et al. 2009). The Nampt gene is directly regulated by the transcriptional activity of CLOCK:BMAL1. SIRT1, a histone deacetylase (HDAC), counteracts the activity of CLOCK:BMAL1, whereas inhibition of NAMPT alleviates this repression. Hence, the redox state has an impact on the activity of transcriptional activators and the rhythmic expression of genes by means of histone modifications. Furthermore, the DNA binding of transcriptional activator complexes is controlled by the redox state. The decreased status of the redox cofactors NAD(H) and NADP(H) increases the transcriptional activity of CLOCK:BMAL1 and NPAS2:BMAL1 complexes (Rutter et al. 2001). AMP-activated protein kinase (AMPK) exhibits diurnal oscillations in its activity. AMPK phosphorylates CRY1, resulting in the destabilization of CRY1 and the release of repression on circadian transcription (Lamia et al. 2009). The significance lies in the fact that the cellular cAMP signaling status has a direct impact on the functioning of the circadian clock. Additionally, the functioning of the clock is reliant on cellular signalling pathways such as cAMP and calcium (O'Neill 2014).

A study by Duong and Weitz (2014), found that mouse PER protein complexes have a histone methyltransferase activity derived from the proteins HP1γ-Suv39h1-2 (suppressor of variegation 3-9 homolog 1 and 2). PERs recruit the HP1γ-Suv39h complex to the Per1 and Per2 promoters, where they play a role in the periodic, di- and tri-methylation of histone H3K9. Suppressing the activity of this complex results in enhanced expression of CLOCK:BMAL1 targets through derepression, leading to a shortened circadian period. The cyclic process of histone acetylation and methylation, facilitated by HDAC1-mediated deacetylation of H3K9, plays a crucial role in feedback repression and the functioning of the circadian clock (Ripperger and Schibler 2006). Multiple other histone methyltransferases play a crucial role in regulating circadian rhythms. The impact of histone modifications on gene expression might vary depending on the specific histone residues involved. These changes can either suppress or stimulate the expression of genes.

MLL1 is responsible for the methylation of lysine residue 4 (K4) on histone H3 and is also found in CLOCK: BMAL1 complexes. MLL1 is responsible for the generation of rhythms in the "activation mark" and the presence of H3K4 trimethylation allows for rhythmic gene expression (Katada and Sassone-Corsi 2010). In a study (Valekunja et al. 2013), it was found that there is a rhythmic occurrence of H3K4 trimethylation at several sites. This process is necessary for the rhythmicity of gene expression and is regulated by Mixed lineage leukemia 3 (MLL3). Importantly, this activity is not influenced by MLL1, CLOCK, or BMAL1. Therefore, a combination of many proteins and biochemical pathways work together to produce circadian rhythms in gene expression by means of histone methylation. WDR5 was shown to be a protein that interacts with PER1 and also has an affinity for PER2 (Brown et al. 2005). WDR5 plays a crucial role as a component of a histone methyltransferase complex, namely in the process of H3K4 trimethylation and H3K9 dimethylation at the Rev-erb-α promoter. Decreasing the levels of WDR5 leads to a reduction in histone methylation at PER1-regulated promoters, but it does not have an effect on the length of the circadian period (Brown et al. 2005).

JARID1a, a histone lysine demethylase, enhances CLOCK-BMAL1 activity and impacts circadian clock operation (DiTacchio et al. 2011). JMJD5, also known as KDM8, is a Jumonj domain protein with histone demethylase activity. It plays a crucial role in regulating the circadian period in both plants and mammalian cells. The depletion of JMJD5 results in a reduction of the circadian period (Jones et al. 2019). The process of adding a methyl group to the arginine residue 3 (H4R3) of histone H4, carried out by the protein arginine methyltransferase 5 (PRMT5), is a modification of the histone that is believed to play a significant role in the control of Per1 gene expression (Na et al. 2012).

Within mammalian cells, the interactions between CRY1 and PRMT5 seem to enhance the process of histone H4R3 dimethylation. Conversely, decreasing the expression of PRMT5 leads to a reduction in H4R3 dimethylation and has an impact on the expression of Per1 (Na et al. 2012). Methylation of DNA and methylation of histone proteins are two different processes that can both impact gene expression. Extended housing of mice in non-24-h "T" cycles results in "aftereffects" on the duration of locomotor activity rhythms. The reversible nature of both DNA methylation and behavioral impacts was observed over time and could be blocked by infusing a DNA methyltransferase inhibitor into the SCN. The primary focus of histone posttranslational modifications is to regulate gene expression on a daily basis. However, our findings demonstrate that long-term changes in DNA methylation can lead to long-term changes in the circadian clock's plasticity (Azzi et al. 2014).

2.5 Post-transcriptional Regulation of the Circadian Clock

Within the circadian clock mechanism, there is a notable level of post-transcriptional control over gene expression. Recent research has shown that there are more regularly expressed transcripts than can be explained just by rhythmic transcription (Koike et al. 2012; Rey et al. 2011). Hence, numerous mRNA cycles are likely to arise via post-transcriptional processes. MicroRNA-mediated post-transcriptional gene silencing is a technique used to regulate the function of the circadian clock through epigenetic means. MicroRNAs are derived from noncoding genes and function by interacting with transcripts of coding genes to suppress their expression. The absence of Dicer, an essential enzyme for the production of microRNAs from precursor transcripts, alters the configuration of gene expression patterns without causing a loss of rhythmicity (Chen and Dorn 2013; Du et al. 2014). Several regulatory RNAs, including microRNAs and long noncoding RNAs, exhibit rhythmic expression patterns and have a role in the rhythmicity of protein-coding genes (Vollmers et al. 2012). Due to the unique sequence requirements for microRNA interactions with their mRNA targets, a vast array of combinations can occur. Furthermore, fluctuations in the length of the poly-A tail in a transcript can have an impact on both the stability of the transcript and its ability to undergo translation. Certain transcripts exhibit circadian oscillations in their poly-A tail length (Kojima et al. 2012; Robinson et al. 1998). This phenomenon ultimately results in circadian rhythms in protein levels.

Right after transcription, pre-mRNAs undergo splicing to eliminate introns. Various genes have the ability to produce several unique protein isoforms due to the selective inclusion of exons within the pre-mRNA. Even if there are no variations in the coding sequences, gene expression levels can still be affected by alternative splicing. This is because the employment of different promoters and variations in the 5′ untranslated regions can influence gene expression. Alternative splicing can result in a distinct 30 untranslated region, leading to variable polyadenylation signals and varying inclusion of sites for microRNA interaction. The regulation of alternative splicing is essentially governed by the interactions between transcripts and RNA-binding proteins.

Recent investigations have shown that alternative splicing and interactions between RNA-binding proteins play a significant role in circadian oscillation and output processes. The cold-inducible RNA-binding protein, CIRP, plays a crucial role in regulating the speed of the circadian clock by interacting with Clock mRNA and influencing alternative polyadenylation (Liu et al. 2013). LARK, an RNA-binding protein known for its impact on circadian rhythms in Drosophila, stimulates the production of PER1 protein in mammals (Kojima et al. 2012). Nocturnin is an RNA deadenylase enzyme that has been discovered in Xenopus. It is characterized as a gene with a strong rhythmic pattern of expression, with its levels particularly high during night-time. Mammalian nocturnin modulates the length of poly-A tails in output genes, while having no impact on the core clock (Kojima and Green 2015). However, it has a significant role in regulating metabolism. The complexes in mice that contain PER also include the RNA helicases DDX5 and DHX9, the active large subunit of RNA polymerase II (Pol II), and SETX, a helicase that facilitates transcriptional termination (Padmanabhan et al. 2012). The activity of SETX is inhibited by the PER-containing complexes, resulting in a delay in the release of RNA polymerase II from Per and Cry pre-mRNAs that are present inside these complexes. Therefore, the complex containing PER attracts proteins that hinder the process of transcriptional re-initiation. The RNA-binding protein Mettl3 plays a role in RNA methylation, which affects transcript processing. Inhibiting this enzyme leads to an increase in the length of the circadian period (Fustin et al. 2013). The circadian cycle duration in flies is regulated by the products of the twenty-four (tyf) and ataxin-2 genes, which impact the translation of the PER protein (Lim and Allada 2013; Zhang et al. 2014).

2.6 Biochemical Oscillations in the Absence of Transcription

The KaiA, KaiB, and KaiC genes in cyanobacteria form an approximately 24-h rhythmic molecular feedback loop (Johnson and Egli 2014). Surprisingly, the production of these three proteins in a controlled laboratory environment, without the involvement of transcription, resulted in the occurrence of circadian rhythms in the phosphorylation of KaiC (Nakajima et al. 2005). Mammalian cells are capable of metabolic oscillations that last approximately 24 h, even without transcription and translation. Clearly, this poses a challenge in aligning with the fundamental mechanism of the core circadian loop, which operates through a feedback loop involving transcription and translation. O'Neill and colleagues discovered circadian rhythms in peroxiredoxins, which exhibited oscillation between reduced and oxidized states in vitro without the presence of nuclei or transcription (O'Neill and Reddy 2011). The variation in the oxidation state of peroxiredoxin is observed in a diverse range of species, which has generated interest in this oscillation as a primitive metabolic oscillator that may have existed before the development of transcriptional/translational feedback loop mechanisms for circadian rhythms in cyanobacteria, fungi, plants, and animals. One unresolved concern is whether this molecular oscillation, which is not dependent on transcription, interacts with the transcriptional apparatus to have a broader impact on cellular activities.

2.7 Mutant Mouse Models of the Clock System

Targeted disruption of multiple genes involved in the fundamental feedback loop has been achieved in mice, resulting in putative null alleles. These mice have been highly valuable in elucidating the function of particular genes in the circadian process. This research has shown that gene families often have redundant functions, meaning that disrupting numerous members of the family is necessary to observe the strongest circadian phenotype. Double-knockout mice that have both the Clock and Npas2 genes disrupted (Debruyne 2008) or the Cry1 and Cry2 genes disrupted (Vitaterna et al. 1999) or the Per1 and Per2 genes disrupted (Bae et al. 2001) are able to survive but do not have circadian rhythms. On the other hand, mice with only one of these genes disrupted show less pronounced circadian patterns.

Bmal1 seems to deviate from the rule of redundancy of function, as mice with only Bmal1 damage do not exhibit any rhythmic patterns (Bunger et al. 2000). The gene Bmal2 (Mop9), which is closely related, is targeted by the CLOCK:BMAL1 complex. However, the expression of Bmal2 is significantly decreased in mice that lack the BMAL1 protein (Shi et al. 2010). When specifically targeting Bmal1, both genes experience a decrease in expression. When BMAL2 is overexpressed via a promoter that does not depend on CLOCK:BMAL1 for expression, it can restore rhythmicity in cells that lack BMAL1 (Shi et al. 2010). The prevalence of gene redundancy among "circadian clock gene families" has significant implications for evaluating gene function in circadian rhythms. Single knockouts may have limited sensitivity in detecting the functional impact of a gene. Using ethyl-nitroso urea (ENU) for mutagenesis results in the creation of specific point mutations. These mutations can lead to the generation of null alleles (causing loss of function, similar to knockouts), dominant-negative alleles, or gain-of-function alleles. The latter two classes of alleles often produce more robust symptoms compared to the matching null allele (as shown below), because these mutant gene products have the capability to influence the activity of an entire gene family.

An advantage of the knockout strategy is that it allows for precise control over when and where a gene is disrupted by employing conditional alleles (for a detailed analysis, refer to (Birky and Bray 2014). There are clock-related genes that have certain areas surrounded by loxP sites ("floxed"), which makes them susceptible to being disrupted by Cre recombinase ("Cre"). There are transgenic lines of organisms that have been modified to express Cre recombinase in particular tissues and cell types. This enables the targeted disruption of gene function in those specific tissues. The Weitz and Bradfield laboratory has created floxed alleles of Bmal1 (Storch et al. 2007; Westgate et al. 2008). The conditional Bmal1 lines have been utilized in various studies to investigate the role of BMAL1 in different organs and tissues, including the vasculature, retina, liver, pancreas, kidney, adipocytes, pituitary gonadotropes, ovary, and brain. Deletions in the Clock gene have been created (Debruyne 2008) and employed in investigations of the liver (DeBruyne et al. 2014). Additionally, researchers have created and examined a variant of CK1δ and two variants of CK1ε that are conditional alleles (Etchegaray et al. 2009; Meng et al. 2008). Furthermore, researchers have successfully created transgenic lines that allow for the regulation of CLOCK, CLOCKΔ19, or BMAL1 expression using doxycycline (Hong et al. 2007; McDearmon et al. 2006). It is crucial to note that these "circadian clock genes" are not solely engaged in circadian rhythms. The phenotype exhibited in a circadian mutant mouse may be attributed to the gene's pleiotropic effects rather than just the absence of circadian clock function.

Several spontaneously occurring mutant alleles of various circadian clock genes have been discovered. The semidominant point mutations consist of mutations in the genes that encode human PER2 (Toh et al. 2001), human CK1δ (Xu et al. 2005), and hamster CK1ε (Lowrey et al. 2000). Furthermore, there have been reports of two novel strains of hamsters exhibiting circadian period characteristics (Monecke et al. 2011). The specific mutated genes have yet to be determined. The use of induced mutagenesis, namely by exposing male mice to ENU and subsequent behavioral screening of the resulting offspring, has resulted in the identification of numerous mutations that impact the function of the circadian clock (Bacon et al. 2004; Vitaterna et al. 1999). The most renowned mutant resulting from mutagenesis is found in a gene known as "circadian locomotor output cycles kaput", commonly referred to as Clock. The affected gene and its mutation were found by positional cloning and transgenic rescue techniques (Antoch et al. 1997; King et al. 1997). The mutant allele has an intronic nucleotide change that causes exon 19 to be excluded throughout the splicing process.

The mutant variant, CLOCKΔ19, maintains the ability to connect with BMAL1 and bind to DNA but does not possess the transcriptional activator function of the normal CLOCK protein (Gekakis et al. 1998). When studied alongside chromosomal deletions, including Clock, the mutant allele demonstrated that the CLOCKΔ19 protein functions as a dominant negative. It actively disrupts the normal function of the wild-type gene product in heterozygotes (King et al. 1997). To clarify, the ClockΔ19 mutation is not a non-functional variant. Experiments involving the removal of the Clock protein in mice (known as "knockout" mice) demonstrate that these animals are still able to maintain consistent daily behavioral patterns, albeit their circadian period is slightly shorter in conditions of constant darkness (DD), and they respond differently to light (Dallmann et al. 2012). In a screening process to identify dominant mutations that impact circadian rhythms in the descendants of males exposed to mutagens, it is highly probable that a null allele of the Clock gene would not have been detected. For instance, individuals with a Clock+/− genotype would not exhibit any noticeable physical or behavioral characteristics. This screen specifically identified the ClockΔ19 mutation. Following mutagenesis tests have been specifically intended to identify recessive mutations or null alleles. However, it remains uncertain whether the slight alteration in the period length of Clock−/− mice in constant darkness would have been detectable even with this screening approach. Therefore, the process of ENU mutagenesis produced a new and powerful variant of the Clock gene that acts dominantly and negatively, resulting in a noticeable phenotype in individuals with only one copy of the gene (Siepka et al. 2007).

This discovery highlights the significance of this circadian gene in regulating biological rhythms, which may have been underestimated if a knockdown approach had been employed. This prompts inquiries on the potential of extensive knockout initiatives to find genes relevant to circadian rhythms, considering the observed trend that gene family members often exhibit redundancy. Anafi et al. (2014) employed a variety of parameters to find potential circadian clock genes, thereby enhancing the ability to detect genes that may have more nuanced yet significant roles. Mutagenesis is an ongoing process that uncovers new insights into the molecular mechanisms underlying the function of the circadian clock. In addition, a mutation in the Fbxl3 gene has been discovered as the cause of the "overtime" phenotype (Siepka et al. 2007). Similarly, a null allele of the Cry1 gene has been identified as the cause of the "part-time" phenotype. Furthermore, a mutation in the Fbxl21 gene has been identified as the cause of the "past-time" phenotype (Yoo et al. 2013). Further mutants have been produced using a comparable screening process conducted at Harwell, UK (Bacon et al. 2004). The Harwell screen discovered that the Afterhours mutation is an additional form of the Fbxl3 gene (Godinho et al. 2007). Additionally, it led to the identification of the gene zinc finger homeobox 3 (Zfhx3) as the one mutated in the Short-circuit (Sci) mutant line, as previously discussed (Parsons et al. 2014). It is still to be confirmed if the other mutants revealed in these screens are novel mutations in known clock genes or new clock genes. The existence of a large number of unidentified mutants that impact circadian rhythms, along with the ongoing discovery of new mechanisms that regulate circadian clocks (such as LHX1, ZFHX1, and CHRONO), suggests that there is a possibility for substantial and continuous advancement in our comprehension of circadian clock mechanisms.

2.8 Conclusion

Undoubtedly, the SCN "connectome" or "wiring diagram" is highly intricate. The suprachiasmatic nucleus (SCN) does not exclusively handle environmental light signals received from the retina, contrary to what is commonly illustrated in a simplistic eskinogram. Conversely, it translates synaptic inputs from numerous distinct brain regions and hormonal signals from the body's outside regions. The SCN stands out from other brain circuits due to its ability to communicate with the entire brain and body through both (poly)synaptic projections across the nervous system and non-synaptic signals to distant cells, organs, and tissues. The suprachiasmatic nucleus (SCN) serves as a critical hub in an intricate and ever-changing circadian network. It combines external information, such as light intensity, with internal information about an animal's arousal, motivation, hormone levels, and more. The integrated circadian timing signal is transmitted to synchronize and coordinate daily patterns in behaviour and physiology.

Significantly, these specific SCN target regions possess distinct molecular clocks and, within the brain, "neuronal clocks" that regulate daily variations in action potential frequency. Hence, clocks in various SCN target regions employ distinct methods to encode circadian time to control their rhythmic outputs. The SCN connectome must encompass not only the inputs and outputs of the SCN but also the reciprocal interaction between these inputs and outputs and the local clocks. An in-depth comprehension of circadian rhythm generation in mammals is crucial for assessing the potential detrimental impact on human health caused by the disturbance of these circuits. Additionally, the intracellular mechanism responsible for circadian rhythmicity is a feedback loop that includes transcription and translation processes. This feedback loop is functional in numerous cell types and enables localized modulation of rhythmic function. Rhythmic transcription is a single element; however, rhythmicity is achieved with a 24-h periodicity through metabolites, chromatin remodelling, and posttranscriptional regulatory processes. Although circadian oscillators function independently inside individual cells, intercellular communication, facilitated by neuropeptides, is crucial for coordinating the activity of these oscillators within the suprachiasmatic nucleus (SCN). This communication helps tie the oscillators together, creating a strong and effective pacemaker that regulates both physiology and behaviour.

References

Abe M, Herzog ED, Yamazaki S, Straume M, Tei H, Sakaki Y, Menaker M, Block GD (2002) Circadian rhythms in isolated brain regions. J Neurosci 22:1

Abrahamson EE, Moore RY (2001) Suprachiasmatic nucleus in the mouse: retinal innervation, intrinsic organization and efferent projections. Brain Res 916:1–2. https://doi.org/10.1016/S0006-8993(01)02890-6

Acosta-Galvan G, Yi C-X, van der Vliet J, Jhamandas JH, Panula P, Angeles-Castellanos M, del Carmen Basualdo M, Escobar C, Buijs RM (2011) Interaction between hypothalamic dorsomedial nucleus and the suprachiasmatic nucleus determines intensity of food anticipatory behavior. Proc Natl Acad Sci 108:14

Akhtar RA, Reddy AB, Maywood ES, Clayton JD, King VM, Smith AG, Gant TW, Hastings MH, Kyriacou CP (2002) Circadian cycling of the mouse liver transcriptome, as revealed by cDNA microarray, is driven by the suprachiasmatic nucleus. Curr Biol 12:7

Alenghat T, Meyers K, Mullican SE, Leitner K, Adeniji-Adele A, Avila J, Bućan M, Ahima RS, Kaestner KH, Lazar MA (2008) Nuclear receptor corepressor and histone deacetylase 3 govern circadian metabolic physiology. Nature 456:7224

Anafi RC, Lee Y, Sato TK, Venkataraman A, Ramanathan C, Kavakli IH, Hughes ME, Baggs JE, Growe J, Liu AC, Kim J, Hogenesch JB (2014) Machine learning helps identify CHRONO as a circadian clock component. PLoS Biol 12:4. https://doi.org/10.1371/journal.pbio.1001840

Anand SN, Maywood ES, Chesham JE, Joynson G, Banks GT, Hastings MH, Nolan PM (2013) Distinct and separable roles for endogenous CRY1 and CRY2 within the circadian molecular clockwork of the suprachiasmatic nucleus, as revealed by the Fbxl3 ^{Afh} mutation. J Neurosci 33:17. https://doi.org/10.1523/JNEUROSCI.4950-12.2013

Annayev Y, Adar S, Chiou Y-Y, Lieb JD, Sancar A, Ye R (2014) Gene model 129 (Gm129) encodes a novel transcriptional repressor that modulates circadian gene expression. J Biol Chem 289:8. https://doi.org/10.1074/jbc.M113.534651

Antoch MP, Song E-J, Chang A-M, Vitaterna MH, Zhao Y, Wilsbacher LD, Sangoram AM, King DP, Pinto LH, Takahashi JS (1997) Functional identification of the mouse circadian Clock gene by transgenic BAC rescue. Cell 89:4

Aschoff J (1960) Exogenous and endogenous components in circadian rhythms. Cold Spring Harb Symp Quant Biol 25:11–28

Asher G, Reinke H, Altmeyer M, Gutierrez-Arcelus M, Hottiger MO, Schibler U (2010) Poly (ADP-ribose) polymerase 1 participates in the phase entrainment of circadian clocks to feeding. Cell 142:6

Atkinson G, Edwards B, Reilly T, Waterhouse J (2007) Exercise as a synchroniser of human circadian rhythms: an update and discussion of the methodological problems. Eur J Appl Physiol 99:331–341

Aton SJ, Herzog ED (2005) Come together, right… now: Synchronization of rhythms in a mammalian circadian clock. Neuron 48:4

Azzi A, Dallmann R, Casserly A, Rehrauer H, Patrignani A, Maier B, Kramer A, Brown SA (2014) Circadian behavior is light-reprogrammed by plastic DNA methylation. Nat Neurosci 17:3

Bacon Y, Ooi A, Kerr S, Shaw-Andrews L, Winchester L, Breeds S, Tymoska-Lalanne Z, Clay J, Greenfield A, Nolan P (2004) Screening for novel ENU-induced rhythm, entrainment and activity mutants. Genes Brain Behav 3:4

Bae K, Jin X, Maywood ES, Hastings MH, Reppert SM, Weaver DR (2001) Differential functions of mPer1, mPer2, and mPer3 in the SCN circadian clock. Neuron 30:2

Balsalobre A, Damiola F, Schibler U (1998) A serum shock induces circadian gene expression in mammalian tissue culture cells. Cell 93:6

Balsalobre A, Brown SA, Marcacci L, Tronche F, Kellendonk C, Reichardt HM, Schutz G, Schibler U (2000) Resetting of circadian time in peripheral tissues by glucocorticoid signaling. Science 289:5488

Barnes JW, Tischkau SA, Barnes JA, Mitchell JW, Burgoon PW, Hickok JR, Gillette MU (2003) Requirement of mammalian Timeless for circadian rhythmicity. Science 302(5644):439–442. https://doi.org/10.1126/science.1086593

Baver SB, Pickard GE, Sollars PJ, Pickard GE (2008) Two types of melanopsin retinal ganglion cell differentially innervate the hypothalamic suprachiasmatic nucleus and the olivary pretectal nucleus. Eur J Neurosci 27:7

Bernardis LL (1973) Disruption of diurnal feeding and weight gain cycles in weanling rats by ventromedial and dorsomedial hypothalamic lesions. Physiol Behav 10:5

Berson D (2003) Strange vision: Ganglion cells as circadian photoreceptors. Trends Neurosci 26:6. https://doi.org/10.1016/S0166-2236(03)00130-9

Birky TL, Bray MS (2014) Understanding circadian gene function: animal models of tissue-specific circadian disruption. IUBMB Life 66:1

Borbély AA (1982) A two process model of sleep regulation. Hum Neurobiol 1:3

Brancaccio M, Maywood ES, Chesham JE, Loudon ASI, Hastings MH (2013) A Gq-Ca2+ axis controls circuit-level encoding of circadian time in the suprachiasmatic nucleus. Neuron 78:4. https://doi.org/10.1016/j.neuron.2013.03.011

Brown SA, Fleury-Olela F, Nagoshi E, Hauser C, Juge C, Meier CA, Chicheportiche R, Dayer J-M, Albrecht U, Schibler U (2005) The period length of fibroblast circadian gene expression varies widely among human individuals. PLoS Biol 3:10

Bugge A, Feng D, Everett LJ, Briggs ER, Mullican SE, Wang F, Jager J, Lazar MA (2012) Rev-erbα and Rev-erbβ coordinately protect the circadian clock and normal metabolic function. Genes Dev 26:7

Buijs RM, Guzmán Ruiz MA, Méndez Hernández R, Rodríguez Cortés B (2019) The suprachiasmatic nucleus; a responsive clock regulating homeostasis by daily changing the set-points of physiological parameters. Auton Neurosci 218:43–50. https://doi.org/10.1016/j.autneu.2019.02.001

Bunger MK, Wilsbacher LD, Moran SM, Clendenin C, Radcliffe LA, Hogenesch JB, Simon MC, Takahashi JS, Bradfield CA (2000) Mop3 is an essential component of the master circadian pacemaker in mammals. Cell 103:7

Busino L, Bassermann F, Maiolica A, Lee C, Nolan PM, Godinho SIH, Draetta GF, Pagano M (2007) SCF^{Fbxl3} controls the oscillation of the circadian clock by directing the degradation of cryptochrome proteins. Science 316:5826. https://doi.org/10.1126/science.1141194

Can A, Schulze TG, Gould TD (2014) Molecular actions and clinical pharmacogenetics of lithium therapy. Pharmacol Biochem Behav 123:3–16. https://doi.org/10.1016/j.pbb.2014.02.004

Carlezon WA, Duman RS, Nestler EJ (2005) The many faces of CREB. Trends Neurosci 28:8

Chen Y, Dorn GW (2013) PINK1-phosphorylated mitofusin 2 is a parkin receptor for culling damaged mitochondria. Science 340:471–475. https://doi.org/10.1126/science.1231031

Cho J, Lee M, Cochrane CS, Webster CG, Fenton BA, Zhao J, Hong J, Zhou P (2020) Structural basis of the UDP-diacylglucosamine pyrophosphohydrolase LpxH inhibition by sulfonyl piperazine antibiotics. Proc Natl Acad Sci USA 117(8):4109–4116. https://doi.org/10.1073/pnas.1912876117

Chong SYC, Xin L, Ptáček LJ, Fu Y-H (2018) Disorders of sleep and circadian rhythms. In: Handbook of clinical neurology, vol 148. Elsevier, Amsterdam, pp 531–538. https://doi.org/10.1016/B978-0-444-64076-5.00034-X

Christian CA, Moenter SM (2008) Vasoactive intestinal polypeptide can excite gonadotropin-releasing hormone neurons in a manner dependent on estradiol and gated by time of day. Endocrinology 149:6. https://doi.org/10.1210/en.2007-1098

Cohen AA, Milot E, Yong J, Seplaki CL, Fülöp T, Bandeen-Roche K, Fried LP (2013) A novel statistical approach shows evidence for multi-system physiological dysregulation during aging. Mech Ageing Dev 134:110–117

Colwell CS, Michel S, Itri J, Rodriguez W, Tam J, Lelievre V, Hu Z, Liu X, Waschek JA (2003) Disrupted circadian rhythms in VIP- and PHI-deficient mice. Am J Phys 285:5. https://doi.org/10.1152/ajpregu.00200.2003

Conway-Campbell BL, Sarabdjitsingh RA, McKenna MA, Pooley JR, Kershaw YM, Meijer OC, De Kloet ER, Lightman SL (2010) Glucocorticoid ultradian rhythmicity directs cyclical gene pulsing of the clock gene period 1 in rat hippocampus. J Neuroendocrinol 22:10. https://doi.org/10.1111/j.1365-2826.2010.02051.x

Coppari R, Ichinose M, Lee CE, Pullen AE, Kenny CD, McGovern RA, Tang V, Liu SM, Ludwig T, Chua SC (2005) The hypothalamic arcuate nucleus: a key site for mediating leptin's effects on glucose homeostasis and locomotor activity. Cell Metab 1:1

Crosio C, Cermakian N, Allis CD, Sassone-Corsi P (2000) Light induces chromatin modification in cells of the mammalian circadian clock. Nat Neurosci 3:12. https://doi.org/10.1038/81767

Dallmann R, Viola AU, Tarokh L, Cajochen C, Brown SA (2012) The human circadian metabolome. Proc Natl Acad Sci 109:7. https://doi.org/10.1073/pnas.1114410109

Debruyne JP (2008) Oscillating perceptions: the ups and downs of the CLOCK protein in the mouse circadian system. J Genet 87:5. https://doi.org/10.1007/s12041-008-0066-7

DeBruyne JP, Weaver DR, Dallmann R (2014) The hepatic circadian clock modulates xenobiotic metabolism in mice. J Biol Rhythm 29:4

Deery M, Jago L (2009) A framework for work-life balance practices: addressing the needs of the tourism industry. Tour Hosp Res 9:2. https://doi.org/10.1057/thr.2009.4

Deurveilher S, Semba K (2005) Indirect projections from the suprachiasmatic nucleus to major arousal-promoting cell groups in rat: Implications for the circadian control of behavioural state. Neuroscience 130:1. https://doi.org/10.1016/j.neuroscience.2004.08.030

DiTacchio L, Le HD, Vollmers C, Hatori M, Witcher M, Secombe J, Panda S (2011) Histone lysine demethylase JARID1a activates CLOCK-BMAL1 and influences the circadian clock. Science 333:6051

Doi M, Hirayama J, Sassone-Corsi P (2006) Circadian regulator CLOCK is a histone acetyltransferase. Cell 125:3. https://doi.org/10.1016/j.cell.2006.03.033

Du N-H, Arpat AB, De Matos M, Gatfield D (2014) MicroRNAs shape circadian hepatic gene expression on a transcriptome-wide scale. elife 3:e02510. https://doi.org/10.7554/eLife.02510

Duong HA, Weitz CJ (2014) Temporal orchestration of repressive chromatin modifiers by circadian clock period complexes. Nat Struct Mol Biol 21:2

Duong HA, Robles MS, Knutti D, Weitz CJ (2011) A molecular mechanism for circadian clock negative feedback. Science 332:6036. https://doi.org/10.1126/science.1196766

Duvall LB, Taghert PH (2013) E and M circadian pacemaker neurons use different PDF receptor signalosome components in drosophila. J Biol Rhythm 28:4. https://doi.org/10.1177/0748730413497179

Engelen E, Janssens RC, Yagita K, Smits VAJ, Van Der Horst GTJ, Tamanini F (2013) Mammalian TIMELESS is involved in period determination and DNA damage-dependent phase advancing of the circadian clock. PLoS ONE 8:2. https://doi.org/10.1371/journal.pone.0056623

Eskin A (1979) Identification and physiology of circadian pacemakers. Fed Proc 38(12):2570–2572

Etchegaray J-P, Yang X, DeBruyne JP, Peters AH, Weaver DR, Jenuwein T, Reppert SM (2006) The polycomb group protein EZH2 is required for mammalian circadian clock function. J Biol Chem 281:30

Etchegaray J-P, Machida KK, Noton E, Constance CM, Dallmann R, Di Napoli MN, DeBruyne JP, Lambert CM, Yu EA, Reppert SM, Weaver DR (2009) Casein kinase 1 delta regulates the pace of the mammalian circadian clock. Mol Cell Biol 29:14. https://doi.org/10.1128/MCB.00338-09

Faber CL, Deem JD, Phan BA, Doan TP, Ogimoto K, Mirzadeh Z, Schwartz MW, Morton GJ (2021) Leptin receptor neurons in the dorsomedial hypothalamus regulate diurnal patterns of feeding, locomotion, and metabolism. elife 10:e63671. https://doi.org/10.7554/eLife.63671

Feng D, Lazar MA (2012) Clocks, metabolism, and the epigenome. Mol Cell 47:2. https://doi.org/10.1016/j.molcel.2012.06.026
Fernandez DC, Chang Y-T, Hattar S, Chen S-K (2016) Architecture of retinal projections to the central circadian pacemaker. Proc Natl Acad Sci 113:21
Fustin J-M, Doi M, Yamaguchi Y, Hida H, Nishimura S, Yoshida M, Isagawa T, Morioka MS, Kakeya H, Manabe I (2013) RNA-methylation-dependent RNA processing controls the speed of the circadian clock. Cell 155:4
Gachon F (2007) Physiological function of PARbZip circadian clock-controlled transcription factors. Ann Med 39:8. https://doi.org/10.1080/07853890701491034
Gallego M, Virshup DM (2007) Post-translational modifications regulate the ticking of the circadian clock. Nat Rev Mol Cell Biol 8(2):139–148
Gallego D, Hernández P, Clavé P, Jiménez M (2006) $P2Y_1$ receptors mediate inhibitory purinergic neuromuscular transmission in the human colon. Am J Physiol 291:4. https://doi.org/10.1152/ajpgi.00474.2005
Gau D, Lemberger T, Von Gall C, Kretz O, Le Minh N, Gass P, Schmid W, Schibler U, Korf HW, Schütz G (2002) Phosphorylation of CREB Ser142 regulates light-induced phase shifts of the circadian clock. Neuron 34:2
Gekakis N, Staknis D, Nguyen HB, Davis FC, Wilsbacher LD, King DP, Takahashi JS, Weitz CJ (1998) Role of the CLOCK protein in the mammalian circadian mechanism. Science 280:5369
Gerber A, Esnault C, Aubert G, Treisman R, Pralong F, Schibler U (2013) Blood-borne circadian signal stimulates daily oscillations in actin dynamics and SRF activity. Cell 152:3
Glass CK, Saijo K, Winner B, Marchetto MC, Gage FH (2010) Mechanisms underlying inflammation in neurodegeneration. Cell 140:6
Godinho SI, Maywood ES, Shaw L, Tucci V, Barnard AR, Busino L, Pagano M, Kendall R, Quwailid MM, Romero MR (2007) The after-hours mutant reveals a role for Fbxl3 in determining mammalian circadian period. Science 316:5826
Gooley JJ, Lu J, Chou TC, Scammell TE, Saper CB (2001) Melanopsin in cells of origin of the retinohypothalamic tract. Nat Neurosci 4:12. https://doi.org/10.1038/nn768
Goriki A, Hatanaka F, Myung J, Kim JK, Yoritaka T, Tanoue S, Abe T, Kiyonari H, Fujimoto K, Kato Y, Todo T, Matsubara A, Forger D, Takumi T (2014) A novel protein, CHRONO, functions as a core component of the mammalian circadian clock. PLoS Biol 12:4. https://doi.org/10.1371/journal.pbio.1001839
Gotter AL, Nimmakayalu MA, Jalali GR, Hacker AM, Vorstman J, Conforto Duffy D, Medne L, Emanuel BS (2007) A palindrome-driven complex rearrangement of 22q11.2 and 8q24.1 elucidated using novel technologies. Genome Res 17:4. https://doi.org/10.1101/gr.6130907
Grippo RM, Tang Q, Zhang Q, Chadwick SR, Gao Y, Altherr EB, Sipe L, Purohit AM, Purohit NM, Sunkara MD (2020) Dopamine signaling in the suprachiasmatic nucleus enables weight gain associated with hedonic feeding. Curr Biol 30:2
Guilding C, Piggins HD (2007) Challenging the omnipotence of the suprachiasmatic timekeeper: are circadian oscillators present throughout the mammalian brain? Eur J Neurosci 25:11. https://doi.org/10.1111/j.1460-9568.2007.05581.x
Guillaumond F, Becquet D, Boyer B, Bosler O, Delaunay F, Franc JL, François-Bellan AM (2012) DNA microarray analysis and functional profile of pituitary transcriptome under core-clock protein BMAL1 control. Chronobiol Int 29:2. https://doi.org/10.3109/07420528.2011.645707
Guzmán-Ruiz MA, Ramirez-Corona A, Guerrero-Vargas NN, Sabath E, Ramirez-Plascencia OD, Fuentes-Romero R, León-Mercado LA, Basualdo Sigales M, Escobar C, Buijs RM (2015) Role of the suprachiasmatic and arcuate nuclei in diurnal temperature regulation in the rat. J Neurosci 35:46. https://doi.org/10.1523/JNEUROSCI.1449-15.2015
Hablitz LM, Molzof HE, Paul JR, Johnson RL, Gamble KL (2014) Suprachiasmatic nucleus function and circadian entrainment are modulated by G protein-coupled inwardly rectifying (GIRK) channels. J Physiol 592:22. https://doi.org/10.1113/jphysiol.2014.282079
Hannibal J (2010) Input pathways to the biological clock. In: Seasonal affective disorder: practice and research. Oxford Biology, Oxford, pp 13–28

Hatanaka F, Matsubara C, Myung J, Yoritaka T, Kamimura N, Tsutsumi S, Kanai A, Suzuki Y, Sassone-Corsi P, Aburatani H (2010) Genome-wide profiling of the core clock protein BMAL1 targets reveals a strict relationship with metabolism. Mol Cell Biol 30:24

Hattar S, Kumar M, Park A, Tong P, Tung J, Yau K-W, Berson DM (2006) Central projections of melanopsin-expressing retinal ganglion cells in the mouse. J Comp Neurol 497(3):326–349. https://doi.org/10.1002/cne.20970

Hirano A, Yumimoto K, Tsunematsu R, Matsumoto M, Oyama M, Kozuka-Hata H, Nakagawa T, Lanjakornsiripan D, Nakayama KI, Fukada Y (2013) FBXL21 regulates oscillation of the circadian clock through ubiquitination and stabilization of cryptochromes. Cell 152:5. https://doi.org/10.1016/j.cell.2013.01.054

Hirata H, Yoshiura S, Ohtsuka T, Bessho Y, Harada T, Yoshikawa K, Kageyama R (2002) Oscillatory expression of the bHLH factor Hes1 regulated by a negative feedback loop. Science 298:5594. https://doi.org/10.1126/science.1074560

Hirayama J, Sahar S, Grimaldi B, Tamaru T, Takamatsu K, Nakahata Y, Sassone-Corsi P (2007) CLOCKmediated acetylation of BMAL1 controls circadian function. Nature 450(7172):1086–1090

Hirota T, Lee JW, Lewis WG, Zhang EE, Breton G, Liu X, Garcia M, Peters EC, Etchegaray J-P, Traver D (2010) High-throughput chemical screen identifies a novel potent modulator of cellular circadian rhythms and reveals CKIα as a clock regulatory kinase. PLoS Biol 8(12):e1000559

Hong H-K, Chong JL, Song W, Song EJ, Jyawook AA, Schook AC, Ko CH, Takahashi JS (2007) Inducible and reversible clock gene expression in brain using the tTA system for the study of circadian behavior. PLoS Genet 3:2

Hosoda H, Kato K, Asano H, Ito M, Kato H, Iwamoto T, Suzuki A, Masushige S, Kida S (2009) CBP/p300 is a cell type-specific modulator of CLOCK/BMAL1-mediated transcription. Mol Brain 2:1. https://doi.org/10.1186/1756-6606-2-34

Hughes AT, Fahey B, Cutler DJ, Coogan AN, Piggins HD (2004) Aberrant gating of photic input to the suprachiasmatic circadian pacemaker of mice lacking the $VPAC_2$ receptor. J Neurosci 24(14):3522–3526. https://doi.org/10.1523/JNEUROSCI.5345-03.2004

Iitaka C, Miyazaki K, Akaike T, Ishida N (2005) A role for glycogen synthase kinase-3β in the mammalian circadian clock. J Biol Chem 280:33. https://doi.org/10.1074/jbc.M503526200

Inouye ST, Kawamura H (1979) Persistence of circadian rhythmicity in a mammalian hypothalamic 'island' containing the suprachiasmatic nucleus. Proc Natl Acad Sci USA 76:11. https://doi.org/10.1073/pnas.76.11.5962

Itri JN, Michel S, Vansteensel MJ, Meijer JH, Colwell CS (2005) Fast delayed rectifier potassium current is required for circadian neural activity. Nat Neurosci 8:5

Jasper MS, Engeland WC (1994) Splanchnic neural activity modulates ultradian and circadian rhythms in adrenocortical secretion in awake rats. Neuroendocrinology 59:2. https://doi.org/10.1159/000126645

Jin X, Shearman LP, Weaver DR, Zylka MJ, De Vries GJ, Reppert SM (1999) A molecular mechanism regulating rhythmic output from the suprachiasmatic circadian clock. Cell 96:1

Johnson CH, Egli M (2014) Metabolic compensation and circadian resilience in prokaryotic cyanobacteria. Annu Rev Biochem 83:221–247

Jones SE, Lane JM, Wood AR, van Hees VT, Tyrrell J, Beaumont RN, Jeffries AR, Dashti HS, Hillsdon M, Ruth KS (2019) Genome-wide association analyses of chronotype in 697,828 individuals provides insights into circadian rhythms. Nat Commun 10(1):343

Jordan SD, Lamia KA (2013) AMPK at the crossroads of circadian clocks and metabolism. Mol Cell Endocrinol 366:2

Kalsbeek A, Teclemariam-Mesbah R, Péavet P (1993) Efferent projections of the suprachiasmatic nucleus in the golden hamster (Mesocricetus auratus). J Comp Neurol 332:3. https://doi.org/10.1002/cne.903320304

Kapfhamer D, Valladares O, Sun Y, Nolan PM, Rux JJ, Arnold SE, Veasey SC, Bućan M (2002) Mutations in Rab3a alter circadian period and homeostatic response to sleep loss in the mouse. Nat Genet 32:2. https://doi.org/10.1038/ng991

Katada S, Sassone-Corsi P (2010) The histone methyltransferase MLL1 permits the oscillation of circadian gene expression. Nat Struct Mol Biol 17:12

Kim M, Suh H, Cho E-J, Buratowski S (2009) Phosphorylation of the yeast Rpb1 C-terminal domain at serines 2, 5, and 7. J Biol Chem 284:39. https://doi.org/10.1074/jbc.M109.028993

King DP, Zhao Y, Sangoram AM, Wilsbacher LD, Tanaka M, Antoch MP, Steeves TD, Vitaterna MH, Kornhauser JM, Lowrey PL (1997) Positional cloning of the mouse circadian clock gene. Cell 89:4

Klein DC, Moore RY, Reppert SM (1991) Suprachiasmatic nucleus: the mind's clock. Oxford University Press, Oxford

Koike N, Yoo S-H, Huang H-C, Kumar V, Lee C, Kim T-K, Takahashi JS (2012) Transcriptional architecture and chromatin landscape of the core circadian clock in mammals. Science 338(6105):349–354. https://doi.org/10.1126/science.1226339

Kojima S, Green CB (2015) Circadian genomics reveal a role for post-transcriptional regulation in mammals. Biochemistry 54:2

Kojima S, Sher-Chen EL, Green CB (2012) Circadian control of mRNA polyadenylation dynamics regulates rhythmic protein expression. Genes Dev 26:2724–2736. https://doi.org/10.1101/gad.208306.112

Kornmann B, Schaad O, Reinke H, Saini C, Schibler U (2007) Regulation of circadian gene expression in liver by systemic signals and hepatocyte oscillators. Cold Spring Harb Symp Quant Biol 72:319–330

Kowalska E, Ripperger JA, Hoegger DC, Bruegger P, Buch T, Birchler T, Mueller A, Albrecht U, Contaldo C, Brown SA (2013) NONO couples the circadian clock to the cell cycle. Proc Natl Acad Sci 110:5. https://doi.org/10.1073/pnas.1213317110

Kudo T, Schroeder A, Loh DH, Kuljis D, Jordan MC, Roos KP, Colwell CS (2011) Dysfunctions in circadian behavior and physiology in mouse models of Huntington's disease. Exp Neurol 228(1):80–90

Kuhlman SJ, McMahon DG (2006) Encoding the ins and outs of circadian pacemaking. J Biol Rhythm 21:6. https://doi.org/10.1177/0748730406294316

Lamia KA, Sachdeva UM, DiTacchio L, Williams EC, Alvarez JG, Egan DF, Vasquez DS, Juguilon H, Panda S, Shaw RJ, Thompson CB, Evans RM (2009) AMPK regulates the circadian clock by cryptochrome phosphorylation and degradation. Science 326(5951):437–440. https://doi.org/10.1126/science.1172156

Lande-Diner L, Stewart-Ornstein J, Weitz CJ, Lahav G (2015) Single-cell analysis of circadian dynamics in tissue explants. Mol Biol Cell 26:22. https://doi.org/10.1091/mbc.E15-06-0403

Langel JL, Smale L, Esquiva G, Hannibal J (2015) Central melanopsin projections in the diurnal rodent, Arvicanthis niloticus. Front Neuroanat 9:93. https://doi.org/10.3389/fnana.2015.00093

Leak RK, Card JP, Moore RY (1999) Suprachiasmatic pacemaker organization analyzed by viral transynaptic transport. Brain Res 819:1–2

Lee Y, Lee J, Kwon I, Nakajima Y, Ohmiya Y, Son GH, Lee KH, Kim K (2010) Coactivation of the CLOCK–BMAL1 complex by CBP mediates resetting of the circadian clock. J Cell Sci 123:20. https://doi.org/10.1242/jcs.070300

Li Y, Liu Z (2016) Coupling mechanism in the gate and oscillator model of the SCN. Physica A 457:62–72

Liao Y, Rust MJ (2021) The circadian clock ensures successful DNA replication in cyanobacteria. Proc Natl Acad Sci 118:20. https://doi.org/10.1073/pnas.2022516118

Lim C, Allada R (2013) Emerging roles for post-transcriptional regulation in circadian clocks. Nat Neurosci 16:11

Liu Y, Hu W, Murakawa Y, Yin J, Wang G, Landthaler M, Yan J (2013) Cold-induced RNA-binding proteins regulate circadian gene expression by controlling alternative polyadenylation. Sci Rep 3:1

Lowrey PL, Shimomura K, Antoch MP, Yamazaki S, Zemenides PD, Ralph MR, Menaker M, Takahashi JS (2000) Positional syntenic cloning and functional characterization of the mammalian circadian mutation tau. Science 288:5465. https://doi.org/10.1126/science.288.5465.483

Marmorstein R, Trievel RC (2009) Histone modifying enzymes: Structures, mechanisms, and specificities. Biochim Biophys Acta 1789:1. https://doi.org/10.1016/j.bbagrm.2008.07.009

Masri S, Sassone-Corsi P (2013) The circadian clock: a framework linking metabolism, epigenetics and neuronal function. Nat Rev Neurosci 14:1

Maywood ES, Fraenkel E, McAllister CJ, Wood N, Reddy AB, Hastings MH, Morton AJ (2010) Disruption of peripheral circadian timekeeping in a mouse model of Huntington's disease and its restoration by temporally scheduled feeding. J Neurosci 30:10199–10204

Mazuski C, Chen SP, Herzog ED (2020) Different roles for VIP neurons in the neonatal and adult suprachiasmatic nucleus. J Biol Rhythm 35:5. https://doi.org/10.1177/0748730420932073

McDearmon EL, Patel KN, Ko CH, Walisser JA, Schook AC, Chong JL, Wilsbacher LD, Song EJ, Hong H-K, Bradfield CA (2006) Dissecting the functions of the mammalian clock protein BMAL1 by tissue-specific rescue in mice. Science 314:5803

Meijer JH, Colwell CS, Rohling JHT, Houben T, Michel S (2012) Dynamic neuronal network organization of the circadian clock and possible deterioration in disease. In: Progress in brain research, vol 199. Elsevier, Amsterdam, pp 143–162. https://doi.org/10.1016/B978-0-444-59427-3.00009-5

Menaker M, Murphy ZC, Sellix MT (2013) Central control of peripheral circadian oscillators. Curr Opin Neurobiol 23:5. https://doi.org/10.1016/j.conb.2013.03.003

Meng QJ, McMaster A, Beesley S, Lu WQ, Gibbs J, Parks D, Collins J, Farrow S, Donn R, Ray D (2008) Ligand modulation of REV-ERBα function resets the peripheral circadian clock in a phasic manner. J Cell Sci 121(21):3629–3635

Meredith P, Strong J, Feeney JA (2006) Adult attachment, anxiety, and pain self-efficacy as predictors of pain intensity and disability. Pain 123:1. https://doi.org/10.1016/j.pain.2006.02.025

Michel S, Itri J, Han JH, Gniotczynski K, Colwell CS (2006) Regulation of glutamatergic signalling by PACAP in the mammalian suprachiasmatic nucleus. BMC Neurosci 7:1–11

Mitsui S, Yamaguchi S, Matsuo T, Ishida Y, Okamura H (2001) Antagonistic role of E4BP4 and PAR proteins in the circadian oscillatory mechanism. Genes Dev 15:8. https://doi.org/10.1101/gad.873501

Moga MM, Moore RY (1997) Organization of neural inputs to the suprachiasmatic nucleus in the rat. J Comp Neurol 389:3

Mohawk JA, Takahashi JS (2011) Cell autonomy and synchrony of suprachiasmatic nucleus circadian oscillators. Trends Neurosci 34:7

Monecke S, Coombs G, Shore AC, Coleman DC, Akpaka P, Borg M, Chow H, Ip M, Jatzwauk L, Jonas D, Kadlec K, Kearns A, Laurent F, O'Brien FG, Pearson J, Ruppelt A, Schwarz S, Scicluna E, Slickers P, Ehricht R (2011) A field guide to pandemic, epidemic and sporadic clones of methicillin-resistant Staphylococcus aureus. PLoS ONE 6:4. https://doi.org/10.1371/journal.pone.0017936

Morin LP (2012) Circadian visual system of mammals. In: How animals see the world. Oxford University Press, Oxford, pp 389–415

Morin F, Bengio Y (2005) Hierarchical probabilistic neural network language model, pp 246–252

Morin LP, Blanchard JH (2001) Neuromodulator content of hamster intergeniculate leaflet neurons and their projection to the suprachiasmatic nucleus or visual midbrain. J Comp Neurol 437(1):79–90

Na J, Lee K, Kim H-G, Shin J-Y, Na W, Jeong H, Lee J-W, Cho S, Kim W-S, Ju B-G (2012) Role of type II protein arginine methyltransferase 5 in the regulation of Circadian Per1 gene. PLoS One 7(10):10

Nader N, Chrousos GP, Kino T (2010) Interactions of the circadian CLOCK system and the HPA axis. Trends Endocrinol Metab 21(5):277–286

Nakahata Y, Sahar S, Astarita G, Kaluzova M, Sassone-Corsi P (2009) Circadian control of the NAD^+ salvage pathway by CLOCK-SIRT1. Science 324(5927):654–657. https://doi.org/10.1126/science.1170803

Nakajima M, Imai K, Ito H, Nishiwaki T, Murayama Y, Iwasaki H, Oyama T, Kondo T (2005) Reconstitution of circadian oscillation of cyanobacterial KaiC phosphorylation in vitro. Science 308:5720

Ni R-J, Shu Y-M, Li T, Zhou J-N (2021) Whole-brain afferent inputs to the caudate nucleus, putamen, and accumbens nucleus in the tree shrew striatum. Front Neuroanat 15:763298. https://doi.org/10.3389/fnana.2021.763298

O'Neill J (2014) Antimicrobial resistance: tackling a crisis for the health and wealth of nations. Rev Antimicrob Resist 7:58

O'Neill JS, Reddy AB (2011) Circadian clocks in human red blood cells. Nature 469:7331

Ohnishi T, Murata T, Watanabe A, Hida A, Ohba H, Iwayama Y, Mishima K, Gondo Y, Yoshikawa T (2014) Defective craniofacial development and brain function in a mouse model for depletion of intracellular inositol synthesis. J Biol Chem 289(15):10785–10796

Ohsaki K, Oishi K, Kozono Y, Nakayama K, Nakayama KI, Ishida N (2008) The role of β-TrCP1 and β-TrCP2 in circadian rhythm generation by mediating degradation of clock protein PER2. J Biochem 144(5):5. https://doi.org/10.1093/jb/mvn112

Padmanabhan K, Robles MS, Westerling T, Weitz CJ (2012) Feedback regulation of transcriptional termination by the mammalian circadian clock PERIOD complex. Science 337:6094

Panda S, Provencio I, Tu DC, Pires SS, Rollag MD, Castrucci AM, Pletcher MT, Sato TK, Wiltshire T, Andahazy M (2003) Melanopsin is required for non-image-forming photic responses in blind mice. Science 301:5632

Parsons S, Valdez-Ramirez V, Congdon R, Williams S (2014) Contrasting patterns of litterfall seasonality and seasonal changes in litter decomposability in a tropical rainforest region. Biogeosciences 11:18

Partch CL, Green CB, Takahashi JS (2014) Molecular architecture of the mammalian circadian clock. Trends Cell Biol 24(2):2

Paul JR, Johnson RL, Jope RS, Gamble KL (2012) Disruption of circadian rhythmicity and suprachiasmatic action potential frequency in a mouse model with constitutive activation of glycogen synthase kinase 3. Neuroscience 226:1–9. https://doi.org/10.1016/j.neuroscience.2012.08.047

Pennartz CM, Hamstra R, Geurtsen AM (2001) Enhanced NMDA receptor activity in retinal inputs to the rat suprachiasmatic nucleus during the subjective night. J Physiol 532(1):1

Perreau-Lenz S, Pévet P, Buijs RM, Kalsbeek A (2004) The biological clock: the bodyguard of temporal homeostasis. Chronobiol Int 21(1):1. https://doi.org/10.1081/CBI-120027984

Piet R, De Croft S, Liu X, Herbison AE (2015) Electrical properties of kisspeptin neurons and their regulation of GnRH neurons. Front Neuroendocrinol 36:15–27. https://doi.org/10.1016/j.yfrne.2014.05.006

Pittendrigh CS (1960) Circadian rhythms and the circadian organization of living systems. Cold Spring Harb Symp Quant Biol 25:159–184. https://doi.org/10.1101/sqb.1960.025.01.015

Porcu A, Nilsson A, Booreddy S, Barnes SA, Welsh DK, Dulcis D (2022) Seasonal changes in day length induce multisynaptic neurotransmitter switching to regulate hypothalamic network activity and behavior. Sci Adv 8:35. https://doi.org/10.1126/sciadv.abn9867

Preitner N, Damiola F, Zakany J, Duboule D, Albrecht U, Schibler U (2002) The orphan nuclear receptor REV-ERBα controls circadian transcription within the positive limb of the mammalian circadian oscillator. Cell 110:2

Punia S, Rumery KK, Yu EA, Lambert CM, Notkins AL, Weaver DR (2012) Disruption of gene expression rhythms in mice lacking secretory vesicle proteins IA-2 and IA-2β. Am J Phys 303:6. https://doi.org/10.1152/ajpendo.00513.2011

Quintero JE, Kuhlman SJ, McMahon DG (2003) The biological clock nucleus: a multiphasic oscillator network regulated by light. J Neurosci 23(22):8070–8076. https://doi.org/10.1523/JNEUROSCI.23-22-08070.2003

Ralph MR, Foster RG, Davis FC, Menaker M (1990) Transplanted suprachiasmatic nucleus determines circadian period. Science 247:4945. https://doi.org/10.1126/science.2305266

Ramsey KM, Yoshino J, Brace CS, Abrassart D, Kobayashi Y, Marcheva B, Hong H-K, Chong JL, Buhr ED, Lee C, Takahashi JS, Imai S, Bass J (2009) Circadian clock feedback cycle through NAMPT-mediated NAD^+ biosynthesis. Science 324(5927):651–654. https://doi.org/10.1126/science.1171641

Rau R, Carlson EN, Back MD, Barranti M, Gebauer JE, Human LJ, Leising D, Nestler S (2021) What is the structure of perceiver effects? On the importance of global positivity and trait-

specificity across personality domains and judgment contexts. J Pers Soc Psychol 120(3):3. https://doi.org/10.1037/pspp0000278
Reinke H, Asher G (2019) Crosstalk between metabolism and circadian clocks. Nat Rev Mol Cell Biol 20(4):227–241. https://doi.org/10.1038/s41580-018-0096-9
Reischl S, Kramer A (2011) Kinases and phosphatases in the mammalian circadian clock. FEBS Lett 585:10. https://doi.org/10.1016/j.febslet.2011.02.038
Rey G, Cesbron F, Rougemont J, Reinke H, Brunner M, Naef F (2011) Genome-wide and phase-specific DNA-binding rhythms of BMAL1 control circadian output functions in mouse liver. PLoS Biol 9:2. https://doi.org/10.1371/journal.pbio.1000595
Ripperger JA, Merrow M (2011) Perfect timing: epigenetic regulation of the circadian clock. FEBS Lett 585:10
Ripperger JA, Schibler U (2006) Rhythmic CLOCK-BMAL1 binding to multiple E-box motifs drives circadian Dbp transcription and chromatin transitions. Nat Genet 38:3
Robinson MJ, Stippec SA, Goldsmith E, White MA, Cobb MH (1998) A constitutively active and nuclear form of the MAP kinase ERK2 is sufficient for neurite outgrowth and cell transformation. Curr Biol 8:21. https://doi.org/10.1016/S0960-9822(07)00485-X
Robles MS, Humphrey SJ, Mann M (2017) Phosphorylation is a central mechanism for circadian control of metabolism and physiology. Cell Metab 25(1):1. https://doi.org/10.1016/j.cmet.2016.10.004
Rossner MJ, Oster H, Wichert SP, Reinecke L, Wehr MC, Reinecke J, Eichele G, Taneja R, Nave K-A (2008) Disturbed clockwork resetting in sharp-1 and sharp-2 single and double mutant mice. PLoS ONE 3:7. https://doi.org/10.1371/journal.pone.0002762
Rutter J, Reick M, Wu LC, McKnight SL (2001) Regulation of clock and NPAS2 DNA binding by the redox state of NAD cofactors. Science 293:5529
Sahar S, Zocchi L, Kinoshita C, Borrelli E, Sassone-Corsi P (2010) Regulation of BMAL1 protein stability and circadian function by GSK3β-mediated phosphorylation. PLoS ONE 5:1. https://doi.org/10.1371/journal.pone.0008561
Saini C, Liani A, Curie T, Gos P, Kreppel F, Emmenegger Y, Bonacina L, Wolf J-P, Poget Y-A, Franken P, Schibler U (2013) Real-time recording of circadian liver gene expression in freely moving mice reveals the phase-setting behavior of hepatocyte clocks. Genes Dev 27(13):1526–1536. https://doi.org/10.1101/gad.221374.113
Saleeba C, Dempsey B, Le S, Goodchild A, McMullan S (2019) A student's guide to neural circuit tracing. Front Neurosci 13:475587
Sato TK, Panda S, Miraglia LJ, Reyes TM, Rudic RD, McNamara P, Naik KA, FitzGerald GA, Kay SA, Hogenesch JB (2004) A functional genomics strategy reveals Rora as a component of the mammalian circadian clock. Neuron 43(4):527–537
Scammell TE, Arrigoni E, Lipton JO (2017) Neural circuitry of wakefulness and sleep. Neuron 93(4):4. https://doi.org/10.1016/j.neuron.2017.01.014
Schmutz I, Wendt S, Schnell A, Kramer A, Mansuy IM, Albrecht U (2011) Protein phosphatase 1 (PP1) is a post-translational regulator of the mammalian circadian clock. PLoS ONE 6:6. https://doi.org/10.1371/journal.pone.0021325
Schneider K, Köcher T, Andersin T, Kurzchalia T, Schibler U, Gatfield D (2012) CAVIN-3 regulates circadian period length and PER: CRY protein abundance and interactions. EMBO Rep 13:12. https://doi.org/10.1038/embor.2012.158
Schoonderwoerd RA, de Rover M, Janse JA, Hirschler L, Willemse CR, Scholten L, Klop I, van Berloo S, van Osch MJ, Swaab DF (2022) The photobiology of the human circadian clock. Proc Natl Acad Sci 119:13
Shearman LP, Sriram S, Weaver DR, Maywood ES, Chaves I, Zheng B, Kume K, Lee CC (2000) Interacting molecular loops in the mammalian circadian clock. Science 288:5468
Shi S, Hida A, McGuinness OP, Wasserman DH, Yamazaki S, Johnson CH (2010) Circadian clock gene Bmal1 is not essential; functional replacement with its paralog, Bmal2. Curr Biol 20:4
Shibata S, Oomura Y, Kita H, Hattori K (1982) Circadian rhythmic changes of neuronal activity in the suprachiasmatic nucleus of the rat hypothalamic slice. Brain Res 247:1. https://doi.org/10.1016/0006-8993(82)91041-1

Shimomura K, Lowrey PL, Vitaterna MH, Buhr ED, Kumar V, Hanna P, Omura C, Izumo M, Low SS, Barrett RK, LaRue SI, Green CB, Takahashi JS (2010) Genetic suppression of the circadian clock mutation by the melatonin biosynthesis pathway. Proc Natl Acad Sci 107:18. https://doi.org/10.1073/pnas.1004368107

Siepka SM, Yoo S-H, Park J, Lee C, Takahashi JS (2007) Genetics and neurobiology of circadian clocks in mammals. Cold Spring Harb Symp Quant Biol 72:1. https://doi.org/10.1101/sqb.2007.72.052

Sonoda T, Li JY, Hayes NW, Chan JC, Okabe Y, Belin S, Nawabi H, Schmidt TM (2020) A noncanonical inhibitory circuit dampens behavioral sensitivity to light. Science 368:6490

Spengler ML, Kuropatwinski KK, Comas M, Gasparian AV, Fedtsova N, Gleiberman AS, Gitlin II, Artemicheva NM, Deluca KA, Gudkov AV, Antoch MP (2012) Core circadian protein CLOCK is a positive regulator of NF-κB–mediated transcription. Proc Natl Acad Sci 109:37. https://doi.org/10.1073/pnas.1206274109

Storch K-F, Paz C, Signorovitch J, Raviola E, Pawlyk B, Li T, Weitz C (2007) Physiological importance of a circadian clock outside the suprachiasmatic nucleus. Cold Spring Harb Symp Quant Biol 72:307–318

Strecker GJ, Wuarin J-P, Dudek FE (1997) GABAA-mediated local synaptic pathways connect neurons in the rat suprachiasmatic nucleus. J Neurophysiol 78:4

Sunkin SM, Ng L, Lau C, Dolbeare T, Gilbert TL, Thompson CL, Hawrylycz M, Dang C (2012) Allen brain atlas: an integrated spatio-temporal portal for exploring the central nervous system. Nucleic Acids Res 41:D1. https://doi.org/10.1093/nar/gks1042

Tahara Y, Hirao A, Moriya T, Kudo T, Shibata S (2010) Effects of medial hypothalamic lesions on feeding-induced entrainment of locomotor activity and liver Per2 expression in Per2: Luc mice. J Biol Rhythm 25:1

Tischkau SA, Mitchell JW, Tyan S-H, Buchanan GF, Gillette MU (2003) Ca2+/cAMP response element-binding protein (CREB)-dependent activation of Per1 is required for light-induced signaling in the suprachiasmatic nucleus circadian clock. J Biol Chem 278:2

Todd WD, Fenselau H, Wang JL, Zhang R, Machado NL, Venner A, Broadhurst RY, Kaur S, Lynagh T, Olson DP, Lowell BB, Fuller PM, Saper CB (2018) A hypothalamic circuit for the circadian control of aggression. Nat Neurosci 21:5. https://doi.org/10.1038/s41593-018-0126-0

Toh KL, Jones CR, He Y, Eide EJ, Hinz WA, Virshup DM, Ptácek LJ, Fu Y-H (2001) An h Per2 phosphorylation site mutation in familial advanced sleep phase syndrome. Science 291:5506

Tosini G, Davidson AJ, Fukuhara C, Kasamatsu M, Castanon-Cervantes O (2007) Localization of a circadian clock in mammalian photoreceptors. FASEB J 21:14

Ueda HR, Hayashi S, Chen W, Sano M, Machida M, Shigeyoshi Y, Iino M, Hashimoto S (2005) System-level identification of transcriptional circuits underlying mammalian circadian clocks. Nat Genet 37:2

Valekunja UK, Edgar RS, Oklejewicz M, van der Horst GT, O'Neill JS, Tamanini F, Turner DJ, Reddy AB (2013) Histone methyltransferase MLL3 contributes to genome-scale circadian transcription. Proc Natl Acad Sci 110:4

Vitaterna MH, Selby CP, Todo T, Niwa H, Thompson C, Fruechte EM, Hitomi K, Thresher RJ, Ishikawa T, Miyazaki J (1999) Differential regulation of mammalian period genes and circadian rhythmicity by cryptochromes 1 and 2. Proc Natl Acad Sci 96(21):12114–12119

Vollmers C, Schmitz RJ, Nathanson J, Yeo G, Ecker JR, Panda S (2012) Circadian oscillations of protein-coding and regulatory RNAs in a highly dynamic mammalian liver epigenome. Cell Metab 16:6. https://doi.org/10.1016/j.cmet.2012.11.004

Vujovic N, Gooley JJ, Jhou TC, Saper CB (2015) Projections from the subparaventricular zone define four channels of output from the circadian timing system. J Comp Neurol 523:18. https://doi.org/10.1002/cne.23812

Wakamatsu H, Yoshinobu Y, Aida R, Moriya T, Akiyama M, Shibata S (2001) Restricted-feeding-induced anticipatory activity rhythm is associated with a phase-shift of the expression of mPer1 and mPer2 mRNA in the cerebral cortex and hippocampus but not in the suprachiasmatic nucleus of mice. Eur J Neurosci 13:6

Walton KM, Fisher K, Rubitski D, Marconi M, Meng Q-J, Sladek M, Adams J, Bass M, Chandrasekaran R, Butler T (2009) Selective inhibition of casein kinase 1ϵ minimally alters circadian clock period. J Pharmacol Exp Ther 330(2):430–439

Watts AG, Swanson LW, Sanchez-Watts G (1987) Efferent projections of the suprachiasmatic nucleus: I. Studies using anterograde transport of Phaseolus vulgaris leucoagglutinin in the rat. J Comp Neurol 258:2. https://doi.org/10.1002/cne.902580204

Weaver DR (1998) The suprachiasmatic nucleus: a 25-year retrospective. J Biol Rhythm 13:2. https://doi.org/10.1177/074873098128999952

Weaver DR, Emery P (2013) Circadian timekeeping. In: Fundamental neuroscience. Elsevier, Amsterdam, pp 819–845

Webb IC, Coolen LM, Lehman MN (2013) NMDA and PACAP receptor signaling interact to mediate retinal-induced scn cellular rhythmicity in the absence of light. PLoS One 8:10

Westgate EJ, Cheng Y, Reilly DF, Price TS, Walisser JA, Bradfield CA, FitzGerald GA (2008) Genetic components of the circadian clock regulate thrombogenesis in vivo. Circulation 117:16

Wiater MF, Mukherjee S, Li A-J, Dinh TT, Rooney EM, Simasko SM, Ritter S (2011) Circadian integration of sleep-wake and feeding requires NPY receptor-expressing neurons in the mediobasal hypothalamus. Am J Physiol 301:5

Willison LD, Kudo T, Loh DH, Kuljis D, Colwell CS (2013) Circadian dysfunction may be a key component of the non-motor symptoms of Parkinson's disease: insights from a transgenic mouse model. Exp Neurol 243:57–66

Wollnik F, Brysch W, Uhlmann E, Gillardon F, Bravo R, Zimmermann M, Schlingensiepen K, Herdegen T (1995) Block of c-Fos and JunB expression by antisense oligonucleotides inhibits light-induced-phase shifts of the mammalian circadian clock. Eur J Neurosci 7:3

Wood S, Loudon A (2014) Clocks for all seasons: unwinding the roles and mechanisms of circadian and interval timers in the hypothalamus and pituitary. J Endocrinol 222:2

Xu Y, Padiath QS, Shapiro RE, Jones CR, Wu SC, Saigoh N, Saigoh K, Ptáček LJ, Fu Y-H (2005) Functional consequences of a CKIδ mutation causing familial advanced sleep phase syndrome. Nature 434:7033

Yamaguchi S, Mitsui S, Yan L, Yagita K, Miyake S, Okamura H (2000) Role of DBP in the circadian oscillatory mechanism. Mol Cell Biol 20:13. https://doi.org/10.1128/MCB.20.13.4773-4781.2000

Yamazaki S, Numano R, Abe M, Hida A, Takahashi R, Ueda M, Block GD, Sakaki Y, Menaker M, Tei H (2000) Resetting central and peripheral circadian oscillators in transgenic rats. Science 288:5466

Yin H-Q, Kim M, Kim J-H, Kong G, Kang K-S, Kim H-L, Yoon B-I, Lee M-O, Lee B-H (2007) Differential gene expression and lipid metabolism in fatty liver induced by acute ethanol treatment in mice. Toxicol Appl Pharmacol 223:3

Yoo S-H, Mohawk JA, Siepka SM, Shan Y, Huh SK, Hong H-K, Kornblum I, Kumar V, Koike N, Xu M (2013) Competing E3 ubiquitin ligases govern circadian periodicity by degradation of CRY in nucleus and cytoplasm. Cell 152(5):1091–1105

Zhang R, Lahens NF, Ballance HI, Hughes ME, Hogenesch JB (2014) A circadian gene expression atlas in mammals: Implications for biology and medicine. Proc Natl Acad Sci 111(45):16219–16224

Zhao Y, Ye L, Liu H, Xia Q, Zhang Y, Yang X, Wang K (2010) Vanadium compounds induced mitochondria permeability transition pore (PTP) opening related to oxidative stress. J Inorg Biochem 104(4):371–378. https://doi.org/10.1016/j.jinorgbio.2009.11.007

Zheng B, Albrecht U, Kaasik K, Sage M, Lu W, Vaishnav S, Li Q, Sun ZS, Eichele G, Bradley A (2001) Nonredundant roles of the mPer1 and mPer2 genes in the mammalian circadian clock. Cell 105(5):683–694

Zhou Q, Liu C, Chen T, Liu Y, Cao R, Ni X, Yang WZ, Shen Q, Sun H, Shen WL (2022) Cooling-activated dorsomedial hypothalamic BDNF neurons control cold defense in mice. J Neurochem 163:3. https://doi.org/10.1111/jnc.15666

Light Sensitivity of the Biological Clock

3

3.1 Introduction

The key to synchronization of short- and or long-term behaviours in organisms so that they happen at the most advantageous period of the day and year has inherent adaptive advantages. An organism will choose the most dependable and highly predictable environmental cues to help it anticipate the "appropriate season" through the ongoing process of natural selection. The annual variation in light-dark (LD) cycle at a given latitude is very predictable, and most organisms take advantage of it. The duration (the length of the light/dark period), the quantity (light intensity), and the quality (spectral composition) of daily light are the three crucial aspects that tell an organism about the time of day and the season. The number of hours of light that an organism perceives is known as the duration. In general, this includes the time between dawn and sunset as well as the twilight hours at each end. The quantity of photons available from a specific light source is referred to as the intensity. The colour, which is typically described as VIBGYOR, is the sensation that results from the activation of a specific class (or classes) of photoreceptors by a light wavelength from the visible light spectrum (380–760 nm). Violet, indigo, and blue are short wavelengths in VIBGYOR, while green and yellow are mid-wavelengths and orange and red are long-wavelengths. Light is available in sinusoidal form in the natural world. The change in intensity and wavelength during twilights, dawn, and dusk is also an important cue for animals, providing them with temporal information.

The neuroendocrine apparatus, collectively referred to as the photoperiodic response system (PRS), interprets light input to produce a reaction to light, known as the photoperiodic response (PR). An input pathway (photoreceptor), a central processor (oscillatory, acts as a clock), and an output pathway (expressed in neuroendocrine events, such as reproduction, body fattening, fur colour, sheep wool growth, etc.) are the three components that make up the PRS. All three PRS components can be found in different tissues in animals. The suprachiasmatic nuclei (SCN, the central oscillatory unit) of the anterior hypothalamus receive light information

S. I. Rizvi, G. Majumdar, *Circadian Clock and Aging*,
https://doi.org/10.1007/978-981-97-6695-6_3

only from the eyes, which are the only photoreceptive organs that do so. Non-mammalian vertebrates may not exhibit this behaviour; birds have many photic input pathways. Through the retina of their lateral eyes, the pineal gland, and the photoreceptors in their brain, they are able to perceive light environments. At each of the three levels, circadian oscillators are present.

3.2 The Photoperiodic Clock

The accuracy with which routine daily and seasonal activities are timed strongly supports the presence of a clock mechanism that is regulated by environmental inputs. According to experimental data, endogenous circa-rhythms synchronised to the time of day (also known as circadian rhythm; circa-about, dies-day) aid in determining how long a day is. These rhythms are undoubtedly light-responsive. As a result, the timekeeping mechanism depends on the photoperiodic entrainment and induction of these light-sensitive rhythms, also known as the *circadian rhythm of photoperiodic photosensitivity* (CRPP).

Under natural lighting conditions (NDL), photoperiodic effects happen either as a result of interaction between two or more independent oscillators brought on by exposure to the LD cycle (internal coincidence) or as a result of coincidence between phases of the photoperiodic oscillator and the LD cycle (external coincidence) (Pittendrigh and Minis 1972). It follows that the two primary aspects of the 24-h daily cycle of illumination, light intensity and wavelength, will have an impact on the photoperiodic entrainment and induction of the CRPP.

The effects of light intensity on photoperiodic induction of gonadal growth and development, a process thought to be mediated by the CRPP, have been the subject of numerous research (Bissonnette and Wadlund 1931; Kumar and Follett 1993). It first showed how light intensity could influence a European starling's reproductive response to day length. This was more than 60 years ago. Since then, researchers have studied the effects of light intensity on photo sexual response in a variety of birds, including house sparrows (Menaker and Eskin 1967; Underwood and Menaker 1970), bobwhite quails (Kirkpatrick 1955), white-crowned sparrows (Farner 1964), domestic ducks (Benoit 1964), domestic turkeys (Nestor and Brown 1972). Most of these studies are those on house sparrows, though. When exposed to light at wavelengths where the photoreceptor(s) mediating the effect are most sensitive or when the photoreceptors are more accessible, a photoperiodic response is maximally elicited.

Photostimulation (weight gain and testis growth) occurred in all groups of black-headed buntings (*Emberiza melanocephala*) exposed to a stimulatory day length of 13L:11D in white, green (528 nm), and red (654 nm) colours at 100 lux intensity, but the responses were noticeably stronger in the red-exposed birds. Other bird species also exhibit comparable reactions (for details, see Kumar and Rani 1999). There have also been reports of wavelength- and intensity-dependent photoperiodic induction in animals. When given to animals during the dark period, a 1-h pulse of green, blue, or near-ultraviolet light, but not red or yellow light, prevented the collapse of the reproductive system in Syrian hamsters. Additionally, different light bandwidths have varied impacts on the collapse of the hamster reproductive system brought on by short photoperiods at equal irradiances (Brainard et al. 1984). Lower vertebrates have also been observed to be affected by the light spectrum. Joshi and Udaykumar (1998) discovered that red light had the biggest effects on the ovarian follicular dynamics in the frog *Ranacyanophlyctis*, followed by yellow and green light. By exposing test subjects to light of various wavelengths at equal irradiance, researchers were able to demonstrate that photoreceptors appear to have precise spectrum sensitivity.

Measuring melatonin, which is regularly produced, at least in mammals, plays a key role in mediating the effects of light on reproduction. This is another way to assess the sensitivity of the photoperiodic clock to light intensity and wavelength. In both LD and constant circumstances (constant darkness [DD] or constant dim light [dimLL]), in vivo and in vitro, plasma melatonin levels are low during the day and high at night, regardless of an animal's nocturnal, diurnal, or crepuscular habits.

Depending on the wavelength of light employed, the effects of light on melatonin secretion at a given irradiance may vary. Albino rats and hamsters both exhibit the highest levels of blue-green's melatonin secretion-inhibitory effects (Brainard et al. 1984; Cardinali et al. 1972). Daylight intensity was found to have some impact on the circadian rhythm of plasma melatonin in a study on European starlings (*Sturnus vulgaris*) (for more information and references, see Kumar et al. 2000). Peak melatonin levels tend to occur earlier in low light (1:0.2 lux) than in bright light (500:0.2 lux), indicating a change in the phasing of the melatonin cycle. A few mammalian species have also been studied for the relationship between the amount of light during the day and/or night and the pattern of plasma melatonin and/or enzymes involved in melatonin synthesis (Griffith and Minton 1992).

It should be remembered, though, that depending on how light is used, it can also have diverse impacts. When administered directly next to or directly on photoreceptors, light in a wide range of wavelengths is stimulatory; yet, when applied away from the photoreceptors, it is not. For instance, when light is directly applied to the brain using a quartz rod, ducks react to both blue and red lights (Benoit 1964). Similar to this, when radioluminescent paint was implanted directly in particular brain locations, Japanese quail (*Coturnixcoturnix japonica*) responded to both blue (455 nm) and orange-yellow (575 nm) light (Banerjee and Chaturvedi 2017). When taken as a whole, all data from spectral investigations on birds support the hypothesis that variations in photoperiodic effects caused by different light wavelengths at a particular intensity are caused by variations in the number of photons received by the photoreceptors (Vriend and Lauber 1973). At the same energy levels, red light emits more photons than green or blue light, and its penetration into brain tissues and subsequent access to the photoreceptors (capture of numerous photons by the photopigments) is much faster and deeper than that of blue or green light, which has shorter wavelengths (Benoit 1964; Foster and Follett 1985). Figure 3.1 depicts the various tissues involved in the generation or entrainment of different kinds of rhythms (daily or annual) and the interrelationship between them. Note that light information in mammals is transmitted to the system only through the eyes.

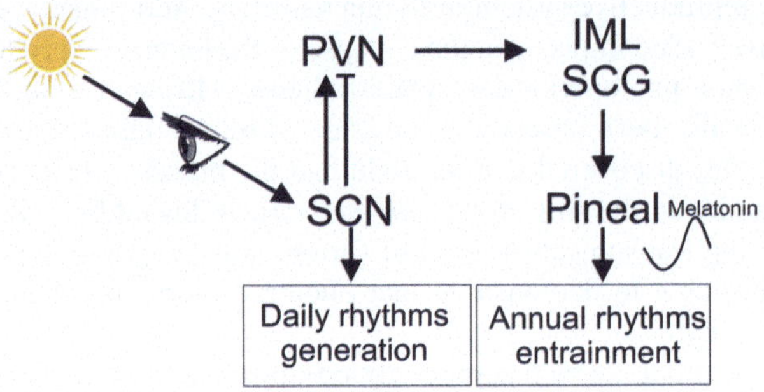

Fig. 3.1 Input of light to various tissues involved in generation or entrainment of different kinds of rhythms (daily or annual) and the interrelationship between them

3.3 CRPP: Phasic Effects

An organism's photoperiodic clock's circadian rhythm can be observed in free running conditions (Kumar and Follett 1993). The process consists of two distinct phases: an entraining phase (subjective day) and an inductive phase (subjective night), which are most effectively illustrated by interrupted-night LD cycles. During these cycles, individuals exposed to short days (e.g., 8L:16D) get a brief light pulse lasting from 1 min to 2 h at night. A photoperiodic response is triggered by the second short light pulse following the initial long light pulse that synchronizes the circadian rhythm of a plant. Light controls the circadian oscillation by influencing the phasing of the oscillation and the photoperiodic induction during the photo-inducible phase in the early subjective night. When light exposure leads to interaction between distinct oscillators (internal coincidence), there is also phase modulation by light. Ironically, a great deal of experimental work has been devoted to evaluating the external coincidence (first proposed by Bunning 1967) or internal coincidence hypotheses, but less attention has been paid to understanding how light influences the phase of the CRPP. The intensity and spectrum sensitivity of the phase-dependent circadian effects of light are still poorly understood.

The entrainment and induction of the biological clock controlling photoperiod-induced gonadal growth and development are demonstrated by indirect evidence from a small number of species. The rate of gonadal growth is extremely slow in migratory bunting (*Emberiza melanocephala* and *Emberiza bruniceps*) kept in a skeleton photo schedule at 100 lux light intensity (Kumar and Rani 1996; Rani and Kumar 2000). In addition, 1 h of night interruption at 850 lux induces considerably larger testes in Japanese quail than exposure to 250 lux (Follett and Milette 1982). Strong evidence supports the idea that light intensity synchronizes long-term seasonal responses that are controlled by clocks. Animals appear to rely on seasonal variations in light intensity to regulate their physiological activity around the equator where annual variation in day duration is too modest to serve as a reliable photoperiodic gauge of the changing season. Using simulations of the natural high- and low-intensity light cycle, this has been experimentally demonstrated in an equatorial species of stonechats (*Saxicola torquata axillaris*) (Gwinner and Scheuerlein

1998). A synchronised annual gonadal cycle under a constant LD cycle with varying light intensities has been seen in these stonechats.

The extremely photosensitive bird models, the black- and redheaded buntings, are being used to test the phasic effects of light intensity and light spectrum on CRPP. Studies used a skeleton paradigm with light pulses that were 6 and 1 h long (6L:6D:1L:11D). These studies have demonstrated that photoperiodic entrainment and induction depended on intensity and spectrum in a number of tests (Kumar and Rani 1996, 1999; Rani and Kumar 2000). White light is more efficient than green light at a given intensity while red light is more efficient than both. In other words, it was found that the stimulation of the avian photoneuroendocrine (PNE) system is less effective with short light wavelengths than with long light wavelengths.

3.4 Induction vs. Entrainment

Differential responses to LD cycles containing either identical entraining and variable inductive pulses or variable entraining and variable inductive pulses indicate wavelength- and intensity-dependent effects on the photoperiodic clock. The possibility of different light intensity thresholds for entraining and inducing light pulses (Kumar and Rani 1996) suggests that entrainment and induction of the CRPP are two distinct photoperiodic phenomena. The threshold light intensity for entrainment is lower than that for induction. For instance, in the migratory redheaded bunting, a 5 lux red light may be efficacious as an entraining agent but not as an inducing agent; the same holds true for a 20 lux white light (Rani and Kumar 2000). In any case, it may be much simpler to entrain the circadian system than to convert photoinductive effects into neuroendocrine events, such as lipogenesis and gametogenesis. Uncertain is whether the distinction occurs at the level of the photoperiodic clock, the level of the endocrine system, or both levels. In addition, it is unknown whether photoperiodic entrainment and induction are mediated by distinct classes of photoreceptors. In ring doves (*Streptopeliaroesegrisea*), Saldanhaand collogues (Saldanha et al. 2001) demonstrated the direct innervation of GnRH neurons by deep brain photoreceptors (DBPs). This may indicate that DBPs, which are required and sufficient for the detection of day-length changes that regulate avian reproduction, are not linked to the reproductive axis via the circadian clock. In other words, "the circadian system is not a necessary intermediary between the sensory and reproductive components" in birds (Saldanha et al. 2001). Since photoperiodic birds can detect minute variations in the characteristics of a photoperiod, it is reasonable to conclude that DBPs also serve as photoperiodic clocks.

3.5 Temporal Photosensitivity

Animals' seasonal reactions are influenced by a variety of elements in addition to light intensity and spectrum. If such were the case, physiological responses to light exposure at a specific spectrum and intensity should be the same regardless of the length of the day (short or long). However, it does not happen because the photosensitive phase of the CRPP responds differently to light pulses supplied at various periods. In redheaded buntings *Emberiza bruniceps* Kumar and Rani (1999), it was demonstrated that the duration- and time-dependent effects of the light pulse inserted in the C/Ji on the pace and magnitude of the testicular response. In buntings, greater C/Ji illumination causes faster rates of gonadal growth, but there is a time limit above which the testicular response cannot be further increased. All of this implies that, after the threshold for light intensity is reached, the "appropriate" photoperiod becomes a more significant issue. In other words, a longer light wavelength or higher light intensity cannot make up for a long photoperiod. According to Burger (1949), a photoperiod of 10.5L:13.5D (a non-stimulatory photoperiod) did not enhance spermatogenesis in male European starlings, regardless of an increase or decrease in light intensity. However, it has also been demonstrated that the impression of day duration in starlings is altered by the reduction in light intensity (Bentley et al. 1998). Thus, it's intriguing to consider the notion that photoperiod and light intensity interact. The influence of light wavelengths on the PRS can be compensatory, as demonstrated by Kumar and colleagues (2000). A prolonged 14 h (highly inductive) photoperiod at short wavelengths caused a gonadal response comparable to or somewhat bigger than that produced by a shorter 12 h (weakly inductive) photoperiod at long light wavelengths.

Depending on the type of LD cycle being employed, there may also be differences in the effects of light intensity. A significant discovery (Kumar and Rani 1999) demonstrated that the action of a single long light pulse in the full photoperiod, extending simultaneously to various phases, differs from that of two discretely falling short light pulses in the skeleton photoperiod at various phases of the photoperiodic oscillator. For a period of 21 weeks, Blackheaded Buntings were exposed to 12L:12D and 6L:5D:1L:12D alternately every day or with continual darkness. The rate and amplitude of photoperiod-induced body fattening and gonadal growth varied significantly between the two photoperiods, indicating distinct differential responses.

3.6 Non-reproductive Circadian Functions

We have a clear understanding of how light affects circadian functions beyond reproduction. Most of the time, the impact of light intensity on circadian activity cycles illustrates the importance of light intensity in synchronizing the biological clock. Free-running is a word in chronobiology that describes the manifestation of a rhythm with a consistent period in unchanging circumstances. The intensity could still influence this specified period, represented by the Greek letter tau. Aschoff's Rule (1981)states that in LL, as light intensity increases, the length of time T increases for nocturnal species and decreases for diurnal species. The phrase "Circadian Rule" pertains to Aschoff's Rule with added considerations of light intensity on the circadian system, particularly its impact on activity (α) and rest (ρ) in the daily activity rhythm. Therefore, diurnal animals experience a decrease in tau, an increase in activity level, and a higher active-to-passive phase ratio in constant bright light conditions as the light intensity rises. Nocturnal species experience contrasting effects. This trend is consistently observed in a great majority of vertebrate groups, including fish, reptiles, amphibians, birds, and nocturnal mammals. The rule is not as effective with most invertebrates, and curiously, it does not appear to be valid for nocturnal mammals, many of which exhibit behaviour similar to diurnal mammals.

Chlamydomonas, a freshwater alga (Kondo et al. 1991), and *Gonyaulax*, a marine alga (Roenneberg and Deng 1997), are two instances of how even organisms with simple organizational structures employ spectral information in photoperiodic timing. Numerous data point to the spectrum effects of light on the circadian clock in vertebrates. For instance, blue light increments are more effective than blue decrements, and yellow decrements are more effective than yellow increments in increasing the activity rhythm of evening-active wild rabbits (Nuboer et al. 1983). The circadian system of the hamster is most sensitive to light wavelengths in the vicinity of 500 nm (Takahashi et al. 1984). While photoreceptors that mediate advance phase changes are most sensitive to light at a wavelength of 520 nm, those that mediate delay phase shifts are most sensitive at a wavelength of 430 nm.

In humans as well, nighttime light may have had a significant impact on the cycles of core temperature (Dijk et al. 1989). Additionally, 1 h of monochromatic light at 509 nm at different intensities inhibits melatonin secretion (Brainard et al. 1984). There isn't much comprehensive research on light's spectrum effects on human circadian rhythms of core temperature and melatonin release. But as demonstrated by a series of fluorescent light trials (Morita et al. 1997), long light wavelengths (such as low colour temperatures of 3000 K and red light) had no impact on people's biological rhythms even at the same light intensity. Mid-wavelength lights, such as green and blue light and light with a high colour temperature of 6500 K, had a stronger impact. The core temperature increased in the morning and decreased at night when exposed to green or blue light. This is because these light colours prevented melatonin secretion from the pineal body from rising at night and lowered it in the morning (Morita et al. 1997).

Despite some compelling results, further research is still needed to fully understand the spectrum sensitivity of the vertebrate circadian clock. For instance, the mouse's circadian system is sensitive to both green light and near-UV irradiance (Provencio and Foster 1995), but we do not understand how the mammalian circadian system makes use of both spectrum signals. Figure 3.2 shows the schematic representation of several non-reproductive functions regulated by light or the circadian clock. Note that the clock is now proven to control mood and cognition in humans and many animal model systems.

3.7 Non-circadian Functions

The effects of light intensity and wavelength have been observed on various physiological functions in addition to behavioural and reproductive rhythms. According to Scott and Siopes (1994), the spectrum of light intensity impacts in birds includes

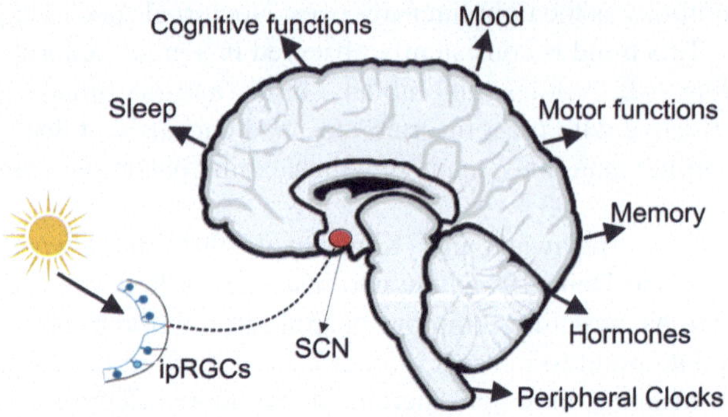

Fig. 3.2 Schematic representation of several non-reproductive functions regulated by light or the circadian clock

those on food and feeding, immune system, and general health. The effects of light wavelengths on magnetoreception in bird migrants, such as Tasmanian silver eyes, *Zosterops l. lateralis*, have been studied in great detail, too (Wiltschko and Wiltschko 2005). According to their research, these migrants were properly oriented in their intended migration direction when exposed to "white" (full spectrum), blue, and green light. Still, they became disoriented when exposed to red light at 633 nm. Other vertebrates, such as fishes (Quinn 1980), amphibians (Lohmann 1991), reptiles (Phillips 1996), and mammals (Marhold and Nagel 1995), have also been reported to exhibit light-dependent compass orientation. However, their wavelength dependence appears to be different from that of birds. Bird growth is also influenced by light intensity and wavelength (Osol et al. 1980). A colour-dependent effect of light was identified on the cellular and humoral immune responses, but not on stress state, in large white turkey chickens at the age of 30 weeks when exposed to blue, green, and red or incandescent light equalized at a photon output (Scott and Siopes 1994). The body temperature, kidney, adrenal, and hypophysis functions, as well as body temperature, have all been linked to light colour in birds (Hollwich and Hollwich 1979).

3.8 Diversity in the Circadian Response to Melatonin in Mammals

Historically, it was believed that nonphotic changes in an animal's physiology and behaviour were insignificant and that the light-dark (LD) cycle was essential for appropriate entrainment processes in mammals. It is now well accepted that a variety of endogenous and centrally acting medications, as well as nonphotic alterations in an animal's environment, may have an impact on the circadian clock. Exogenous melatonin may function as an effective time cue for mammalian circadian systems, according to mounting research over the past two decades. Early research in rats showed that timed melatonin administration could advance free-running activity rhythms, entrain and phase them, and influence the rate and direction of re-entrainment for light-dark cycle phase-shifts (Pévet 2003). These studies led to the initial hypothesis that the mammalian circadian system was sensitive to melatonin only during a small window of time, close to the end of the subjective day. There may be additional times of the day when the circadian clock is susceptible to the hormone, according to more recent research on mice, palm squirrels, and humans (Gnocchi and Bruscalupi 2017). The responses to melatonin across species appear to be very diverse and complex. Given this, it is remarkable that entrainment experiments have only been reported in a small number of species and that just four different mammals' melatonin phase response curves have been documented. Although this option has not been fully explored, it is likely that melatonin affects diurnal and nocturnal species differently.

3.8.1 Entrainment to Melatonin

The timing of melatonin injection is crucial for entraining circadian rhythms. When the beginning of the active phase coincides with the time of administration, short bursts of melatonin administered intravenously, orally (Redman et al. 1983), or intravenously (Slotten et al. 1999), will entrain rat activity rhythms. In contrast, no entrainment takes place when rats have constant access to melatonin-containing drinking water (Armstrong 1989). Recently, it was examined if the length of melatonin administration affected entrainment (Pitrosky et al. 1999). All rats were melatonin-dependent after 1 and 8 h. However, only 50% of rats became melatonin-dependent after 16 h. As the infusion's duration lengthened, it was discovered that the phase angle of entrainment, which was calculated using activity onset and infusion onset as the phase markers, became increasingly negative. 13/15 male rats that displayed rhythm disruption under constant light (LL, 250–300 lux) after being subjected to an unusual lighting regime were able to synchronise or partially synchronise their activity cycles with daily melatonin injections (Chesworth et al. 1987). These results were seen as evidence in favour of the idea that melatonin can affect the coupling between oscillators that drive activity rhythm, and they suggest that melatonin might work better in rats with broken rhythms than in rats with intact rhythms under LL.

In contrast to the earlier study done with male Long-Evans rats (Marumoto et al. 1996) observed that melatonin injections entrain intact free-running rhythms of rats housed in dim LL (3 lux) but did not synchronise disturbed activity rhythms of male Sprague-Dawley rats kept in bright LL (300 lux). The disparity in results between these two investigations may be due to strain differences, the varied lighting regimes, or the different recording equipment utilised to measure activity (wheel-running vs. general locomotor activity).

Two control rats demonstrated susceptibility to the injection method in the first investigation that described melatonin entrainment in rats (Redman et al. 1983). One of the rats became accustomed to receiving saline injections, and the other displayed synchronisation over a period of days. Therefore, it has been hypothesised that melatonin's effects might be mediated by modifications in overall arousal levels (Mrosovsky 1997). Redman and Roberts (Redman 1997) investigated this idea by eliminating wheel-running in melatonin-injected mice. They discovered that entrainment continued even in these circumstances. Additionally, there was no impact on entrainment when rats were immobilised for 3 h following melatonin injections (Marumoto et al. 1996). Therefore, it is unlikely that changes in activity levels would mediate the melatonin entrainment mechanism.

When given to the foetus through the mother with lesioned SCN or to the newborn directly up until postnatal day 6 in Syrian hamsters, melatonin injections entrain activity rhythms (Davis and Mannion 1988; Grosse and Hastings 1996). Melatonin has no effect when a single dose is given to adult Syrian hamsters without handling the animals (Hastings et al. 1997). As a result, in Syrian hamsters, sensitivity to the circadian effects of melatonin seems to depend (at least in part) on developmental age.

Various hamster strains may have various melatonin entrainment phases. Melatonin infusions given 6 or 8 h per day entrain the rhythms of the wheel-running activities of adult Syrian and Djungarian hamsters with pinealectomies (Px) kept in low light levels (LL; 0.1–5 lux) (Kirsch et al. 1993). Entrainment happened in Djungarian hamsters when the start of the melatonin infusion coincided with the start of activity (CT12), much like it did in other nocturnal species that had previously been studied. Entrainment, on the other hand, took place in Syrian hamsters when the start of activity started 3–5 h before the start of the infusion period. The difference in endogenous melatonin between species was used to explain this variation. Syrian hamsters produce melatonin 3–5 h after the start of the dark phase, but Djungarian hamsters, like rats, begin producing it at the beginning of the dark phase (Kirsch et al. 1993). Activity starts in both hamster strains around the time when night falls. In light of these findings, it would appear that the phase of the endogenous melatonin cycle rather than the activity rhythm determines the circadian sensitivity to melatonin. The melatonin's entraining abilities in diurnal (non-human) mammals have not received much research.

To diurnal Indian palm squirrels (*Funambulus pennanti*) kept initially under LD cycles and later under LL (14 lux), melatonin in food was given at either zeitgeber time (ZT) 0 (light onset) or ZT12 (dark onset) (Rajaratnam and Redman 1997). At both studied times of the day, it was discovered that squirrel activity rhythms were synchronized to melatonin. One theory for why squirrels respond to melatonin at (at least) two circadian phases, but rats only do so during activity initiation, is based on the species's tau (τ; free running period). Long Evans rats typically free-run with 'τ' >24 h, whereas palm squirrels may do so with 'r' or >24 h. Depending on whether the animal's 'τ' is less than or larger than 24 h, phase delays or phase advances must be induced in order to entrain palm squirrel activity cycles with melatonin. The fact that this species' circadian system responds to melatonin at more than one circadian phase may therefore be explained by the significant degree of variability in 'τ' for this species. Exogenous melatonin does not appear to entrain all animals. It was also discovered that daily melatonin injections did not entrain the free-running activity cycles of Israeli sand rats (*Psammomys obesus*) kept under dim LL (−0.1 lux)

(Rajaratnam and Redman 1997). There certainly seem to be species differences in reaction to melatonin, albeit direct comparisons are challenging due to the diversity in technique.

According to a number of human investigations, melatonin treatment stabilizes some blind and sighted patients' free-running sleep-wake patterns (Lockley et al. 2000). The effects of melatonin on the sleep-wake rhythm in these experiments, however, were not accompanied by effects on endocrine rhythms, suggesting that the circadian system was only partially entrained or that the apparent entrainment was simply a sign of melatonin's sleep-promoting effects.

3.8.2 Melatonin Administration Under LD Entrainment

Long-Evans rats frequently exhibit a negative phase angle difference (PAD) of up to several hours during LD cycles. Activity onset was phased towards dark onset when these rats were treated with melatonin or its agonist S-20098 for either 8 or 9 days (Armstrong and Redman 1991). Melatonin was unable to advance the rhythm when the negative PAD was larger than 3 h, most likely because the injections occurred beyond the melatonin's circadian window of sensitivity. The rhythm of endogenous melatonin was shown to phase advance by 1–3 h in 5/11 patients when daily oral dosage of melatonin was given to human subjects at 1700 h for 3–4 weeks (Arendt 1997). Melatonin medication may have expedited the sleep-wake cycle because respondents self-reported increased evening fatigue and, in some circumstances, evening sleep. Daily melatonin treatment at 1930 h (Tzischinsky et al. 1993) or 2200 h (Dahlitz et al. 1991) accelerated the sleep-wake cycle in individuals with delayed sleep phase syndrome (DSPS). Thus, the evidence from both healthy persons and DSPS patients supports melatonin's phase-advancement effects in rats. The Djungarian hamster (*P. sungorus*) likewise exhibits reactions to timed melatonin injections that are similar to the effects of the hormone on phase advancement in rats. Daily melatonin injections given 4 h before the beginning of the dark phase advanced the onset of activity and prolonged activity in adult Djungarian hamsters maintained under LD 16:8, but only in those animals that displayed physiological responses to melatonin administration (such as gonadal regression, moult, and body weight loss) (Puchalski and Lynch 1988).

Investigations have been done into how exogenous melatonin affects re-entrainment after LD cycle phase shifts. Continual melatonin administration to rats via an implant close to the SCN was found to hasten the re-entrainment of the adrenocortical rhythm following LD cycle reversal (Murakami et al. 1983). Since then, research has demonstrated that the re-entrainment mechanism responds to melatonin differently depending on the phase of administration and the strength of the LD phase shift. With the right timing, melatonin administration can change the direction and speed of re-entrainment. All rats re-entrained by phase advancing when melatonin was given at the old dark onset time following an 8-h phase advance, whereas all control rats (whether vehicle-treated or unhandled) re-entrained by phase delaying (Redman and Armstrong 1988). When melatonin was supplied at the moment of optimum sensitivity for entrainment, that is, during the pre-shift dark onset, melatonin-injected rats re-entrained faster than vehicle-injected controls after a 5 h phase advance. Pineal NAT rhythms of rats given melatonin at the new dark onset time re-entrained more quickly than those of controls after an 8-h phase advance (Illnerov et al. 1989). Other rodents have verified the effects of melatonin on the rate of re-entrainment. After a 6 h LD phase advance or phase delay, respectively (Sharma et al. 1999; Van Reeth et al. 1998).

Melatonin treatment has been shown to lessen the subjective symptoms of jet lag and speed up the re-entrainment of circadian rhythms to actual or simulated transmeridian travel in humans, according to a number of field investigations and laboratory simulations (Arendt 1997). Jet lag symptoms did not significantly differ between treatment groups in a recent large-scale randomised, double-blind trial, which gave 257 subjects travelling from New York to Oslo (6 h eastward) 3 different melatonin regimens or a placebo (Spitzer et al. 1999). The subjects of this study only arrived in New York (originally from Oslo) 4–5 days before their travel back to Oslo, which may have been inadequate time for the circadian system to properly adjust to New York time before the shift, as noted by the authors. As a result, the timing of the melatonin delivery to these participants may have been off (Arendt 1997).

3.8.3 Melatonin Phase Response Curves

Behavioural PRCs to melatonin have been demonstrated in many diverse mammals, including humans (Lewy and Sack 1997), Long Evans rats (Armstrong 1989), and field mice (Sharma et al. 1999). Melatonin injections cause definite and consistent phase advances of activity rhythms in the late subjective day in Long-Evans rats. Humans receiving melatonin intravenously (Zaidan et al. 1994) or orally (Lewy and Sack 1997) have similarly experienced phase-advancing effects. In contrast, the phase advance part in humans is around between CTs 6 and 14. Phase delays to melatonin have also been documented in all species investigated thus far, with the exception of rats. In *M. booduga,* however, a single dose of melatonin dose-dependently triggered phase advances in the rhythm of the locomotor activity (maximal advances CT22) and phase delays (maximal delays CT4). As a result, it does not seem that all animals exhibit the same melatonin PRC in terms of its general shape and time course (Sharma et al. 1999).

3.8.4 Site(s) of Melatonin Action

The SCN may be the locus of melatonin action in the mammalian circadian clock, according to several lines of evidence. Early research revealed that Pinx and enucleation had no discernible impact on melatonin entrainment in rats (Armstrong 1989; Marumoto et al. 1996).

Research using the melatonin analogue 2-[125I]-iodomelatonin discovered high-affinity melatonin-binding sites in the rat SCN (Vaněček 1988). Additionally, fluctuations in the density of melatonin SCN binding sites on a daily basis have been observed (Gauer et al. 1993). Also, 2-deoxyglucose absorption in the rat SCN was modulated by melatonin injections in vivo (Cassone et al. 1988). The rhythm in SCN neuronal activity is phase advanced when, as previously mentioned. A hypothalamic brain slice preparation comprising the SCN is treated with melatonin in vitro (McArthur et al. 1997). Last but not least, melatonin administered in the late subjective day immediately phase advances the SCN rhythm in photic induction of c-fos (Sumová and Illnerová 1996), and melatonin administered in the late subjective night (CT22) induces c-fos expression in the SCN of rats held in DD. Melatonin antagonists have recently been used to explore the role of the melatonin receptor subtypes MTI (Mella) and MT2 (Mellb) in the phase-advancing effects of melatonin, showing that MT2 (Mellb) subtype may be a likely candidate for mediating the circadian effects of melatonin (Hunt et al. 2001) but further evidence is required to support this claim.

3.9 Conclusion

It is crucial to analyse the impact of light on the environment, particularly its effects on the biology of animals, including human health, comfort, and biological rhythms. Laboratory research demonstrates varying effects of light intensity and wavelength on circadian processes in birds, potentially influencing adaptive physiological responses generated by photoperiod. For instance, in diurnal species, the synchronization to light cycles at a low intensity happens early in the day, while the response to higher light intensity occurs later in the day during spring and summer when days are long enough to reach the photosensitive phase of the CRPP. An avian photoperiodic clock may not be highly sensitive to red light but exhibits the greatest reaction. This could be due to the dominance of red light in the avian PRS and its superior ability to penetrate brain tissues in the light environment. Light intensity and spectrum also control certain physiological processes in animals. Species may exhibit significant variety in their responses to different light wavelengths and intensities. The species-specific spectrum and intensity sensitivity of the PRS could have significant ecological effects. Species inhabiting the same environment may develop distinct temporal strategies to enhance their survival and efficiency.

In humans, the fluctuations in core temperature and melatonin levels are closely linked to the feeling of sleepiness. An illuminated pulse at night may impact core body temperature and melatonin levels, thereby disrupting sleep patterns. Empirical evidence indicates that light with a low colour temperature is suitable for low-level nighttime lighting, whereas light with a high colour temperature is recommended for morning use when higher illumination levels are needed. Physical and mental disorders may arise from a mismatch between the biological rhythm phase and the individual's lifestyle, which is influenced by light. This is evident when there is inadequate synchronization to the light-dark cycle caused by either light deficiency or variations in light phases. Examples include jet lag syndrome resulting from swift travel across time zones, sleeplessness among nocturnal workers, and seasonal affective disorder prevalent throughout winter months. One crucial factor in optimizing light usage is synchronizing daily light exposure with biological rhythms. It is crucial to analyse how light at different times of the day affects biological cycles and physiological functioning. It is essential to further investigate how repeated exposure to everyday light with a specific spectral distribution affects biological rhythms and any potential impacts of chromatic adaptation in a living setting.

References

Arendt J (1997) Safety of melatonin in long-term use(?). J Biol Rhythm 12:673–681. https://doi.org/10.1177/074873049701200624

Armstrong SM (1989) Melatonin and circadian control in mammals. Experientia 45:932–938. https://doi.org/10.1007/BF01953050

Armstrong SM, Redman JR (1991) Melatonin: a chronobiotic with anti-aging properties? Med Hypotheses 34:300–309. https://doi.org/10.1016/0306-9877(91)90046-2

Aschoff J (1981) A survey on biological rhythms. Springer, Berlin

Banerjee S, Chaturvedi CM (2017) Testicular atrophy and reproductive quiescence in photorefractory and scotosensitive quail: involvement of hypothalamic deep brain photoreceptors and GnRH-GnIH system. J Photochem Photobiol B 175:254–268. https://doi.org/10.1016/j.jphotobiol.2017.09.005

Benoit J (1964) The role of the eye and of the hypothalamus in the photostimulation of gonads in the duck. Ann N Y Acad Sci 117:204–216. https://doi.org/10.1111/j.1749-6632.1964.tb48175.x

Bentley GE, Goldsmith AR, Dawson A, Briggs C, Pemberton M (1998) Decreased light intensity alters the perception of day length by male European starlings (Sturnus vulgaris). J Biol Rhythm 13:148–158

Bissonnette TH, Wadlund APR (1931) Spermatogenesis in sturnus vulgaris: refractory period and acceleration in relation to wave length and rate of increase of light ration. J Morphol 52:403–427. https://doi.org/10.1002/jmor.1050520204

Brainard GC, Richardson BA, Hurlbut EC, Steinlechner S, Matthews SA, Reiter RJ (1984) The influence of various irradiances of artificial light, twilight, and moonlight on the suppression of pineal melatonin content in the Syrian hamster. J Pineal Res 1:105–119. https://doi.org/10.1111/j.1600-079X.1984.tb00202.x

Bunning E (1967) The physiological clock. Springer, Berlin

Burger JW (1949) A review of experimental investigations on seasonal reproduction in birds. Wilson Bull 61:211–230

Cardinali DP, Larin F, Wurtman RJ (1972) Control of the rat pineal gland by light spectra. Proc Natl Acad Sci 69:2003–2005. https://doi.org/10.1073/pnas.69.8.2003

Cassone VM, Roberts MH, Moore RY (1988) Effects of melatonin on 2-deoxy-[1-14C]glucose uptake within rat suprachiasmatic nucleus. Am J Physiol 255:R332–R337. https://doi.org/10.1152/ajpregu.1988.255.2.R332

Chesworth MJ, Cassone VM, Armstrong SM (1987) Effects of daily melatonin injections on activity rhythms of rats in constant light. Am J Physiol 253:R101–R107. https://doi.org/10.1152/ajpregu.1987.253.1.R101

Dahlitz M, Alvarez B, Parkes JD, English J, Arendt J, Vignau J (1991) Delayed sleep phase syndrome response to melatonin. Lancet 337:1121–1124. https://doi.org/10.1016/0140-6736(91)92787-3

Davis FC, Mannion J (1988) Entrainment of hamster pup circadian rhythms by prenatal melatonin injections to the mother. Am J Physiol 255:R439–R448. https://doi.org/10.1152/ajpregu.1988.255.3.R439

Dijk DJ, Beersma D, Daan S, Lewy AJ (1989) Bright morning light advances the human circadian system without affecting NREM sleep homeostasis. Am J Physiol 256:R106–R111

Farner DS (1964) Time measurement in vertebrate photoperiodism. Am Nat 98:375–386

Follett BK, Milette JJ (1982) Photoperiodism in quail: testicular growth and maintenance under skeleton photoperiods. J Endocrinol 93:83–90. https://doi.org/10.1677/joe.0.0930083

Foster RG, Follett BK (1985) The involvement of a rhodopsin-like photopigment in the photoperiodic response of the Japanese quail. J Comp Physiol A 157:519–528. https://doi.org/10.1007/BF00615153

Gauer F, Masson-Pévet M, Skene DJ, Vivien-Roels B, Pévet P (1993) Daily rhythms of melatonin binding sites in the rat pars tuberalis and suprachiasmatic nuclei; evidence for a regulation of melatonin receptors by melatonin itself. Neuroendocrinology 57:120–126. https://doi.org/10.1159/000126350

Gnocchi D, Bruscalupi G (2017) Circadian rhythms and hormonal homeostasis: pathophysiological implications. Biology 6:10. https://doi.org/10.3390/biology6010010

Griffith MK, Minton JE (1992) Effect of light intensity on circadian profiles of melatonin, prolactin, ACTH, and cortisol in pigs. J Anim Sci 70:492–498. https://doi.org/10.2527/1992.702492x

Grosse J, Hastings MH (1996) A role for the circadian clock of the suprachiasmatic nuclei in the interpretation of serial melatonin signals in the Syrian hamster. J Biol Rhythm 11:317–324. https://doi.org/10.1177/074873049601100405

Gwinner E, Scheuerlein A (1998) Seasonal changes in daylight intensity as a potential zeitgeber of circannual rhythms in equatorial Stonechats. J Ornithol 139:407–412

Hastings MH, Duffield GE, Ebling FJ, Kidd A, Maywood ES, Schurov I (1997) Nonphotic signalling in the suprachiasmatic nucleus. Biol Cell 89:495–503. https://doi.org/10.1016/S0248-4900(98)80005-1

Hollwich F, Hollwich F (1979) Light and metabolic functions. Presented at the influence of ocular light perception on metabolism in man and in animal. Springer, Berlin, pp 55–63

Hunt AE, Al-Ghoul WM, Gillette MU, Dubocovich ML (2001) Activation of MT2 melatonin receptors in rat suprachiasmatic nucleus phase advances the circadian clock. Am J Physiol Cell Physiol 280:C110–C118. https://doi.org/10.1152/ajpcell.2001.280.1.C110

Illnerov H, Trentini G, Maslova L (1989) Melatonin accelerates reentrainment of the circadian rhythm of its own production after an 8-h advance of the light-dark cycle. J Comp Physiol A 166:97–102. https://doi.org/10.1007/BF00190214

Joshi BN, Udaykumar K (1998) Changes in ovarian follicular kinetics in intact blinded and parietal shielded frogs exposed to different spectra of light. Gen Comp Endocrinol 109:310–314. https://doi.org/10.1006/gcen.1997.7034

Kirkpatrick CM (1955) Factors in photoperiodism of bobwhite quail. Physiol Zool 28:255–264

Kirsch R, Belgnaoui S, Gourmelen S, Pévet P (1993) Daily melatonin infusion entrains free-running activity in Syrian and Siberian hamsters. In: Light and biological rhythms in man. Elsevier, Amsterdam, pp 107–120. https://doi.org/10.1016/B978-0-08-042279-4.50011-X

Kondo T, Johnson CH, Hastings JW (1991) Action spectrum for resetting the circadian phototaxis rhythm in the CW15 strain of Chlamydomonas: I. Cells in darkness. Plant Physiol 95:197–205

Kumar V, Follett BK (1993) The circadian nature of melatonin secretion in Japanese quail (*Coturnix coturnix japonica*). J Pineal Res 14:192–200. https://doi.org/10.1111/j.1600-079X.1993.tb00502.x

Kumar V, Rani S (1996) Effects of wavelength and intensity of light in initiation of body fattening and gonadal growth in a migratory bunting under complete and skeleton photoperiods. Physiol Behav 60:625–631. https://doi.org/10.1016/S0031-9384(96)80040-0

Kumar V, Rani S (1999) Light sensitivity of the photoperiodic response system in higher vertebrates: wavelength and intensity effects. Indian J Exp Biol 37:1053–1064

Kumar V, Rani S, Malik S (2000) Wavelength of light mimics the effects of the duration and intensity of a long photoperiod in stimulation of gonadal responses in the male blackheaded bunting (Emberiza melanocephala). Curr Sci 79:508–510

Lewy AJ, Sack RL (1997) Exogenous melatonin's phase-shifting effects on the endogenous melatonin profile in sighted humans: a brief review and critique of the literature. J Biol Rhythm 12:588–594. https://doi.org/10.1177/074873049701200614
Lockley SW, Skene DJ, James K, Thapan K, Wright J, Arendt J (2000) Melatonin administration can entrain the free-running circadian system of blind subjects. J Endocrinol 164:R1–R6. https://doi.org/10.1677/joe.0.164r001
Lohmann KJ (1991) Magnetic orientation by hatchling loggerhead sea turtles (Caretta caretta). J Exp Biol 155:37–49
Marhold S, Nagel A (1995) The energetics of the common mole rat Cryptomyś, a subterranean eusocial rodent from Zambia. J Comp Physiol B 164:636–645
Marumoto N, Murakami N, Katayama T, Kuroda H, Murakami T (1996) Effects of daily injections of melatonin on locomotor activity rhythms in rats maintained under constant bright or dim light. Physiol Behav 60:767–773. https://doi.org/10.1016/0031-9384(96)00100-X
McArthur AJ, Hunt AE, Gillette MU (1997) Melatonin action and signal transduction in the rat suprachiasmatic circadian clock: activation of protein kinase C at dusk and dawn. Endocrinology 138:627–634. https://doi.org/10.1210/endo.138.2.4925
Menaker M, Eskin A (1967) Circadian clock in photoperiodic time measurement: a test of the Bünning hypothesis. Science 157:1182–1185. https://doi.org/10.1126/science.157.3793.1182
Morita T, Tokura H, Wakamura T, Park S-J, Teramoto Y (1997) Effects of the morning irradiation of light with different wavelengths on the behavior of core temperature and melatonin in humans. Appl Hum Sci 16:103–105
Mrosovsky N (1997) Problems in interpreting the effects of drugs on circadian rhythms. In: Redfern PH, Lemmer B (eds) Physiology and pharmacology of biological rhythms. Springer, Berlin, pp 415–434. https://doi.org/10.1007/978-3-662-09355-9_15
Murakami N, Hayafuji C, Sasaki Y, Yamazaki J, Takahashi K (1983) Melatonin accelerates the reentrainment of the circadian adrenocortical rhythm in inverted illumination cycle. Neuroendocrinology 36:385–391. https://doi.org/10.1159/000123485
Nestor KE, Brown KI (1972) Light intensity and reproduction of Turkey hens. Poult Sci 51:117–121. https://doi.org/10.3382/ps.0510117
Nuboer JFW, Van Nuys WM, Van Steenbergen JC (1983) Colour changes in a light regimen as synchronizers of circadian activity. J Comp Physiol 151:359–366. https://doi.org/10.1007/BF00623911
Osol J, Foss D, Carew L Jr (1980) Effects of light environment and pinealectomy on growth and thyroid function in the broiler cockerel. Poult Sci 59:647–653
Pévet P (2003) Melatonin in animal models. Dialogues Clin Neurosci 5:343–352. https://doi.org/10.31887/DCNS.2003.5.4/ppevet
Phillips JB (1996) Magnetic navigation. J Theor Biol 180:309–319
Pitrosky B, Kirsch R, Malan A, Mocaer E, Pevet P (1999) Organization of rat circadian rhythms during daily infusion of melatonin or S20098, a melatonin agonist. Am J Physiol 277:R812–R828. https://doi.org/10.1152/ajpregu.1999.277.3.R812
Pittendrigh CS, Minis DH (1972) Circadian systems: longevity as a function of circadian resonance in Drosophila melanogaster. Proc Natl Acad Sci U S A 69:1537–1539. https://doi.org/10.1073/pnas.69.6.1537
Provencio I, Foster RG (1995) Circadian rhythms in mice can be regulated by photoreceptors with cone-like characteristics. Brain Res 694:183–190
Puchalski W, Lynch R (1988) Daily melatonin injections affect the expression of circadian rhythmicity in Djungarian hamsters kept under a long-day photoperiod. Neuroendocrinology 48:280–286. https://doi.org/10.1159/000125023
Quinn NJ (1980) Analysis of temporal changes in fish assemblages in Serpentine Creek, Queensland. Environ Biol Fishes 5:117–133
Rajaratnam SMW, Redman JR (1997) Effects of daily melatonin administration on circadian activity rhythms in the diurnal Indian palm squirrel (Funambulus pennanti). J Biol Rhythm 12:339–347. https://doi.org/10.1177/074873049701200406

Rani S, Kumar V (2000) Phasic response of the photoperiodic clock to wavelength and intensity of light in the redheaded bunting, Emberiza bruniceps. Physiol Behav 69:277–283. https://doi.org/10.1016/S0031-9384(99)00247-4

Redman JR (1997) Circadian entrainment and phase shifting in mammals with melatonin. J Biol Rhythm 12:581–587

Redman JR, Armstrong SM (1988) Reentrainment of rat circadian activity rhythms: effects of melatonin. J Pineal Res 5:203–215. https://doi.org/10.1111/j.1600-079X.1988.tb00782.x

Redman J, Armstrong S, Ng KT (1983) Free-running activity rhythms in the rat: entrainment by melatonin. Science 219:1089–1091

Roenneberg T, Deng T-S (1997) Photobiology of the Gonyaulax circadian system. I. Different phase response curves for red and blue light. Planta 202:494–501

Saldanha CJ, Silverman A-J, Silver R (2001) Direct innervation of GnRH neurons by encephalic photoreceptors in birds. J Biol Rhythm 16:39–49. https://doi.org/10.1177/074873040101600105

Scott R, Siopes T (1994) Light color: effect on blood cells, immune function and stress status in turkey hens. Comp Biochem Physiol A Physiol 108:161–168

Sharma VK, Singaravel M, Subbaraj R, Chandrashekaran MK (1999) Timely administration of melatonin accelerates reentrainment to phase-shifted light-dark cycles in the field mouse Mus booduga. Chronobiol Int 16:163–170. https://doi.org/10.3109/07420529909019083

Slotten HA, Pitrosky B, Pévet P (1999) Influence of the mode of daily melatonin administration on entrainment of rat circadian rhythms. J Biol Rhythm 14:347–353. https://doi.org/10.1177/074873099129000759

Spitzer RL, Terman M, Williams JBW, Terman JS, Malt UF, Singer F, Lewy AJ (1999) Jet lag: clinical features, validation of a new syndrome-specific scale, and lack of response to melatonin in a randomized, double-blind trial. Am J Psychiatry 156:1392–1396. https://doi.org/10.1176/ajp.156.9.1392

Sumová A, Illnerová H (1996) Melatonin instantaneously resets intrinsic circadian rhythmicity in the rat suprachiasmatic nucleus. Neurosci Lett 218:181–184. https://doi.org/10.1016/S0304-3940(96)13159-1

Takahashi JS, DeCoursey PJ, Bauman L, Menaker M (1984) Spectral sensitivity of a novel photoreceptive system mediating entrainment of mammalian circadian rhythms. Nature 308:186–188

Tzischinsky O, Shlitner A, Lavie P (1993) The association between the nocturnal sleep gate and nocturnal onset of urinary 6-sulfatoxymelatonin. J Biol Rhythm 8:199–209. https://doi.org/10.1177/074873049300800303

Underwood H, Menaker M (1970) Photoperiodically significant photoreception in sparrows: is the retina involved? Science 167:298–301. https://doi.org/10.1126/science.167.3916.298

Van Reeth O, Olivares E, Turek FW, Granjon L, Mocaer E (1998) Resynchronisation of a diurnal rodent circadian clock accelerated by a melatonin agonist. Neuroreport 9:1901–1905. https://doi.org/10.1097/00001756-199806010-00043

Vaněček J (1988) Melatonin binding sites. J Neurochem 51:1436–1440. https://doi.org/10.1111/j.1471-4159.1988.tb01108.x

Vriend J, Lauber JK (1973) Effects of light intensity, wavelength and quanta on gonads and spleen of the deer mouse. Nature 244:37–38. https://doi.org/10.1038/244037a0

Wiltschko W, Wiltschko R (2005) Magnetic orientation and magnetoreception in birds and other animals. J Comp Physiol A 191:675–693

Zaidan R, Geoffriau M, Brun J, Taillard J, Bureau C, Chazot G, Claustrat B (1994) Melatonin is able to influence its secretion in humans: description of a phase-response curve. Neuroendocrinology 60:105–112. https://doi.org/10.1159/000126726

Circadian System and Ageing in Rodent Models

4

4.1 Introduction

Circadian rhythms (CR), or the biological clock, are essential components of almost all living things and can be seen at cellular, behavioural, or physiological levels. In the previous chapters, we discussed how a rhythm is generated in synchrony to the external environment, how the clock operates, and what its regulating factors are. However, as animals age, their internal clocks become less regular and more susceptible to disrupting external stimuli. In this chapter, we will discuss the physiology of the rodent circadian system, rhythms and clock gene disturbance in aging and age-related disorders, and models of genetic (clock gene mutant/knockout mice) and environmental (light-dark regimes modification) circadian disturbance. As we've highlighted, the circadian timekeeping mechanism in mammals is extremely conserved both anatomically and genetically. It is set up hierarchically, with the central pacemaker able to relay information to the periphery oscillators to synchronize its rhythm with outside light cues. The components of the circadian clock at the cellular level are the clock genes' protein products, which form a number of transcription-translation feedback loops. Internal cycles last roughly 24 h, but they are constantly altered by signals from the central pacemaker. The primary element in humoral timekeeping is thought to be melatonin. Circadian rhythmic functions include rest-activity cycles, hormone synthesis, and antioxidant activities. CR are altered as a result of the aging circadian system's inability to accurately manage peripheral rhythms. In addition to sleep loss or fragmentation, age-related deterioration in circadian function can severely worsen life quality. The circadian clock also has a role in cognitive, cardiovascular, and metabolic problems; hence chronodisruption may significantly exacerbate these maladies. According to animal models, age-related illnesses and rapid aging are caused by CR disruption, which can occur due to environmental or hereditary factors. In mammalian organisms, nearly all physiological systems are subject to diurnal fluctuations. It must meet three criteria in order to be considered a "circadian rhythm": (a) have an endogenous period of about 24 h; (b)

S. I. Rizvi, G. Majumdar, *Circadian Clock and Aging*,
https://doi.org/10.1007/978-981-97-6695-6_4

be able to entrain in response to outside stimuli; and (c) display temperature adjustment. The production and secretion of practically all hormones, different metabolic processes, gene expression, body temperature, heart rate, and blood pressure are only a few examples of rhythmic functioning (Bell-Pedersen et al. 2005). Only a few of these, including locomotion, sleep-wake cycles, body temperature, and rhythms of eating and drinking, can, however, be assessed non-invasively (Nakamura et al. 2008). As a result, the majority of CR research focuses on these functions. Actograms, special plots where daily activity patterns are aligned with one another, are used to easily notice and evaluate the behavioural rhythm, which is the most obvious rhythm. Phase shifts may be visible with such a representation (Jud et al. 2005). Constant conditions (LL or DD), phase shift or jet lag, long or short light cycles, and total chronodisruption are among the techniques used to assess CR consolidation.

4.2 Central and Peripheral Oscillators

Every clock, biological or otherwise, has an internal mechanism producing rhythmic ticking, an input system that aligns the clock with astronomical time, and output signals needed for information transfer. The early 1970s saw the discovery of suprachiasmatic nuclei (SCN), a circadian pacemaker; the precise mechanism of light sensitivity was not known until 2002, when specific cells in the retina detecting non-visual photic information were found. Discovering neural connections from the retina to the SCN (retinohypothalamic tract, RHT), it became clear how important the SCN is in producing circadian rhythm (CR) and entrainment to an external light schedule (Honma et al. 2012). Mammals' central clock is found in SCN nuclei, each consisting of roughly 10,000 cells somewhat above the optical chiasm. Fascinatingly, lesions in SCN totally impaired circadian rhythmicity but the transplanting of SCN restored rhythmicity (Li and Satinoff 1998). Silver and colleagues (1996) suggested the existence of humoral components that carry rhythmic impulses since SCN transplantation at least partially recovers CR even if it is placed in a capsule that does not allow axon outgrowth. Among these are examples vaspopressin, prokineticin 2 (PK2), and cardiotrophin-like cytokine (CLC). Corresponding genes are controlled in a circadian way by clock genes since their promoter regions contain E-box. Their harmonic release controls body temperature CRs, hormone synthesis (Li et al. 2012), and rest-activity cycles.

Though every neuron in the SCN may create a 24-h rhythm, the strength of the rhythm depends on synchronising; light is the most effective entrainment signal. Paracrine neuropeptide signals allow neurons in the SCN to coordinate their activity; interruption in the development of synaptic vesicles causes messy neuron firing.

Synchronisation within the SCN has been shown to be produced by vasoactive intestinal peptide (VIP), arginine-vasopressin peptide (AVP), and gastrin-releasing peptide (GRP) (Hastings et al. 2014). Since VIP or VPAC2 (VIP receptor) mutant mice show an arrhythmia, disturbance of this connection causes the loss of rhythmic SCN output.

SCN neurons provide inhibitory signals to pinealocytes in the presence of light; in the absence of light, they stop doing this and let pinealocytes generate melatonin. Melatonin is far more ancient than the pineal gland since it is discovered in unicellular eukaryotes. Synthesised mostly at night in diurnal and nocturnal species from tryptophan, this hormone Its two membrane G-protein linked receptors are MT1 and MT2 (Dubocovich et al. 2010). Melatonin also affects the activation of nuclear receptors (rors and Rev-ERBs), therefore controlling the production of clock genes (Hazlerigg et al. 1996). Apart from axons controlling pineal gland activity, SCN has neuronal projections to the midline thalamus and the hypothalamus. A mammalian organism consists of several peripheral oscillators whose phases could not coincide. Every tissue contains systems for paracrine and cell-to-cell communication to synchronise its functional aspects. Every cell also has a molecular calendar. In this part, the liver will be used as a model of a peripheral oscillator. Highly metabolically active, this tissue serves only to collect nutrition and remove toxins during feeding time. Though food consumption cycles can occasionally be disturbed, generally nutrition corresponds with the activity phase. For those with either nocturnal eating syndrome (NES) or an erratic diet, this is especially true. Using the limited feeding schedule—that which food is given for just 1–2 h per day—this scenario is replicated in laboratory animals. It was also shown that whilst the SCN rhythm is still entrained by light, under such settings, the mouse liver re-entrains its molecular clock according to eating time (Mendoza et al. 2005). The periphery clocks are not strictly regulated by the central clock. Retaining SCN rhythm, liver-specific Rev-Erba knockout animals have a hepatic circadian transcriptome overlapping but not exactly like those of WT mice (Vieira et al. 2014). Peripheral tissues thus have their own rhythm; the SCN coordinates these tissues.

4.3 Circadian System in Aged Rodents

The process of ageing can have an effect on circadian rhythms, resulting in disrupted sleep and activity patterns. Elderly individuals who are in good health may display sleep and wake dysfunctions, such as struggling to start sleeping, experiencing repeated awakenings at night, and waking up early in the morning (Buysse 2005). The changes in the daily sleep and waking patterns that have been observed can be attributed to the interaction between homeostatic sleep mechanisms and circadian clock parameters (Dijk et al. 1989). Furthermore, the process of clock ageing has been found to have an impact on mood and memory, as well as being a contributing factor to age-related cognitive decline (Grandner et al. 2012; Nebes et al. 2009). There is also a positive correlation between reduced amplitude of locomotor activity rhythm and an elevated risk of mortality. In this section, the effect of aging on clock properties is discussed (see Fig. 4.1).

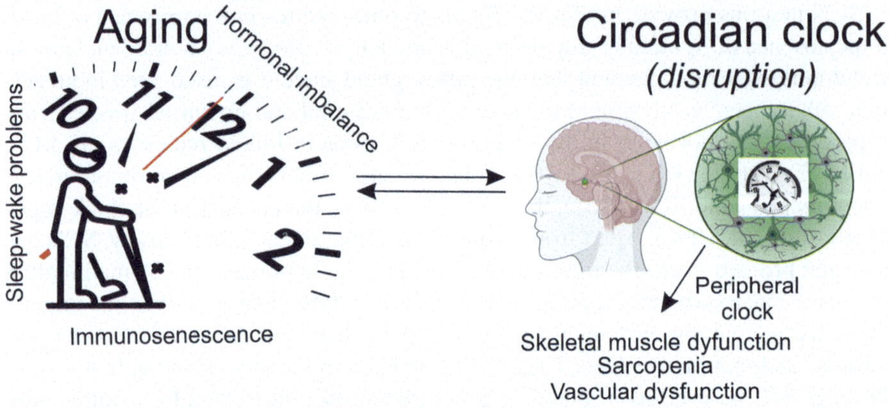

Fig. 4.1 Schematic representation of the effect of aging on clock properties

4.3.1 Circadian Rhythm Disturbance Related to Ageing

Aging disrupts the majority of rhythmic functions of the body and the synchronization of peripheral systems. Animals and humans share certain characteristics of aging rhythms. The rhythmicity of body temperature, the sleep-wake cycle, and locomotor activity are diminished in older rodents (Duncan et al. 2012; Turek et al. 2001).

Mice, rats, and hamsters display reduced intensity and division of their daily activity pattern. In mature animals, the duration of activity is reduced and becomes more fragmented (Nakamura et al. 2011; Turek et al. 2001). The magnitude of the body temperature and corticosterone rhythms also diminished. Interestingly, rats and mice showed individual differences in the disturbance of their daily rhythms. While most older animals had reduced intensity in their sleep-wake cycles, others still had strong circadian rhythms. It is important to acknowledge that different oscillators undergo ageing and loss of rhythm at different rates. Weinert (2000) found that the corticosterone rhythm was weakened before the activity rhythm, whereas the body temperature rhythm remained strong in animals with decreased activity amplitude.

Aged rodents exhibit age-related phase shift. Most research has shown that mice and hamsters tend to display phase-advanced behaviour, which means they have an earlier commencement of rest and activity. However, C57BL mice tend to exhibit phase delay, meaning they have a later onset of rest and activity (Valentinuzzi et al. 1997). Certain species and/or strains have shown a reduction in the duration of their free-running period, as observed by Davidson et al. (2006). This reduction in period length may be a factor contributing to phase-advanced behaviour. According to Davidson et al. (2006), animals become more susceptible to jet lag as they need more circadian cycles to resynchronize.

It is important to comprehend which clock components degrade over time. The aging circadian system loses sensitivity to entraining light cues, but lens transmission in aged golden hamsters does not change appreciably (Zhang et al. 2004). Using rodents devoid of rod and cone photoreceptors, a 40% decrease in the number of retinal ganglion cells in aged animals was also observed (Semo et al. 2003). This decrease may contribute to the diminished circadian photosensitivity observed in aged animals and humans. With age, the ticking of the central timepiece becomes less rhythmic.

4.3.2 The Ageing SCN

The aging SCN neither loses many neurons nor undergoes atrophic alterations despite a decline in SCN functions (Duffy et al. 2015; Farajnia et al. 2015). The in vivo recording of SCN multiunit neural activity in rodents has revealed a decrease in the amplitude of the neuronal firing rhythm. In addition, CR was degraded in the subpara-ventricular zone, one of the SCN's primary outputs (Nakamura et al. 2011).

Electrophysiological recordings of neurons in brain slices containing the suprachiasmatic nucleus (SCN) from elderly rats, mice, and hamsters have identified changes in neuronal activity associated with ageing (Aujard et al. 2001; Satinoff et al. 1991). Research conducted by Farajnia et al. (2015) has provided evidence indicating that neuronal activity in the SCN of older rodents becomes desynchronized. Furthermore, the SCN of aged rats experiences a reduction in the quantity of cells that express VIP and AVP, which are peptides essential for synchronised neuronal function (Roozendaal et al. 1987). Furthermore, a reduction in the expression of SIRT1 in the brains of elderly rodents, coupled with a deficit in this gene, leads to a disruption in the rhythm that resembles the process of ageing. The upregulation of SIRT1 in mice shields them from the deterioration of rhythmic functions and the downregulation of clock genes (CGs) associated with ageing (Chang and Guarente 2013). Research has demonstrated that the transplantation of foetal suprachiasmatic nucleus (SCN) into the third ventricle of elderly animals enhances their circadian rhythmicity (Li and Satinoff 1998). Furthermore, when wild-type SCN was transplanted into arrhythmic Cry1/Cry2 knockout or Clock mutant animals, it resulted in the development of behavioural rhythms even without functional peripheral oscillators (Sujino et al. 2003). Therefore, the rhythmic signals produced by the SCN are crucial for the continuous maintenance of circadian rhythms throughout one's life.

4.3.3 Aging Pineal Gland

In aged animals and humans, pineal melatonin synthesis decreases, resulting in a diminution of the amplitude of the CR. The pineal gland of aged rodents endures minor morphological changes. While some researchers have not documented any age-related alterations in rats, others have observed a decline in the overall count of pinealocytes, an augmentation in the thickness of the connective tissue capsule and the number of connective tissue cells, the presence of calcified concretions, and certain cellular changes at the ultrastructural level (Kunz et al. 1999). Also, it was found that in the aged rat pineal gland, sympathetic axon innervation was reduced (Schmidt et al. 2006).

Loss of axonal projections from SCN carrying information about the light-dark cycle appears to be related to a decrease in the rhythm of pineal melatonin production. Changes in morphology that reflect decreased synthetic and secretory activity may result from diminished innervation. Additionally, it is believed that melatonin signal plays an important function in pinealocytes maintenance (Cernuda-Cernuda et al. 2000) compared to the pineal glands of melatonin-deficient C57BL/6J mice to those of a normal CBA strain at the age of 10 months; the latter strain contained more pinealocytes. During the early stages of development, the pineal glands of both strains were similar in size and characteristics. Typically, the ageing of the circadian system results in the disruption of rhythms at several levels, such as behaviour, tissues, and cells. This may be caused by a decline in the concentration of melatonin.

4.3.4 Aging Molecular Clock

Disruption of synchronised regulation of clock genes can modify the operation of the SCN. Multiple studies demonstrate changes in the expression of clock genes

Table 4.1 Alterations in age-related expression of clock genes in different rodent species

Species	Tissue	Effects on the clock (gene/protein expression)
Mice	SCN	Rhythm loss of Bmal1, rhythmic Cry1 expression
	Liver	Decrease of Per1, Per2, Cry1 gene expression amplitude, clock gene rhythmic expression
	Heart	Decrease of Cry2 gene amplitude, clock gene rhythmic expression
	Brain regions	Changes in phase and/or amplitude of clock and Bmal1 protein synthesis
Rat	SCN	Loss of Bmal1, Cry1, Cry2 rhythm, change of phase of Per1, Per2 gene expression
	PVN, pineal	No changes in Cry1, Per1 or Per2 gene expression rhythm
	Liver	Decrease in evening expression of Per2 gene
	Heart	Reduction of Per2 expression, increase in Bmal1 expression
	Kidney	Change of Per1 gene expression phase
	Dermal fibroblasts	Decrease in Per1 expression
Syrian Hamster	SCN	Changes in gene expression rhythmicity of clock and Bmal1
	Skeletal muscles	No changes found in in Per2 and Bmal1 expression rhythm
	Brain regions	Loss of Bmal1 expression rhythm, decrease in Per2 and Bmal1 gene expression amplitude

in the suprachiasmatic nucleus (SCN), as well as other areas of the brain and peripheral tissues, in aged animals. Please refer to Table 4.1 for more details. The impact of age on clock genes is unclear and may be masked by light-dark cycles that coordinate their activity in the SCN. Old mice kept in a state of permanent darkness exhibited reduced amplitude of Per2 expression and a longer circadian period compared to mice of the same age kept in a regular LD cycle. Furthermore, a study by Nakamura et al. (2015) found that neurons in rodents living in disrupted light-dark cycles did not exhibit synchronisation. The ageing of the circadian system is an intricate process encompassing various behavioural, anatomical, and molecular changes. It also involves both central and peripheral oscillators. Throughout nearly five decades of studying the circadian system, a significant volume of data has been gathered, although no malfunctioning cogwheel has been found in the old clock.

Does aging have a bidirectional relationship with circadian disruption? In the next sections, we will examine the impact of circadian disturbance on age-related changes, considering both environmental and genetic factors. Additionally, we will explore therapies that can modify the circadian system and its relationship with ageing.

4.4 Analysis of the Circadian Rhythms in Rodent Models of Accelerated Ageing and Age-Related Pathologies

Circadian rhythm disturbances associated with ageing are a prevalent phenomenon observed in various species, including humans, and are distinguished by alterations in both behavioural patterns and physiological processes (Brock 1991; Van Gool and Mirmiran 1986). Aged animals exhibit comparable behavioural alterations (Valentinuzzi et al. 1997). Research has demonstrated that deficiencies in object recognition and disturbances in circadian rhythms manifest at comparable stages in animal models, particularly in rodents. Therefore, these model systems have made significant progress in investigating the correlation between clock and ageing. However, further research is necessary in this area due to its significance in developing novel treatments for the aged.

4.4.1 Clock Genes in Rodent Models of Age-Related Pathologies

Neurodegenerative diseases such as Parkinson's disease (PD), Alzheimer's disease, and Huntington's disease have common symptoms, including the gradual death of neurons, cognitive or motor impairments, and disturbances in regions responsible for regulating activity, sleep-wake cycle, and body temperature. Abnormalities in diverse rhythmic functions are also seen in animal models of similar disorders.

Chemicals such as 6-hydroxydopamine, 1-methyl-4-phenyl-1,2,3,6-tetrahydropyridine, and rotenone can cause degeneration in dopaminergic neurons, leading to tremors, similar to what occurs in Parkinson's disease patients. Additionally, genetically modified mice, specifically those that overexpress alpha-synuclein or have the MITOPARK mutation, can also experience spontaneous degeneration of dopaminergic neurons as they age. Both mouse and rat models display weakened patterns of rest-activity, CG expression rhythms, and disturbances in other rhythmic functions (Videnovic and Willis 2016). Administration of levodopa (Boulamery et al. 2010), melatonin (Mattam and Jagota 2015), or ATP (Hayashi et al. 2013) has been demonstrated to possess a neuroprotective impact and partially reinstate rhythmic functions.

Alzheimer's disease in humans is marked by the buildup of protein aggregates composed of s-protein and b-amyloid in the brain tissue. This leads to memory disorders, sleep disruption, alterations in body temperature and activity patterns, and the disruption of melatonin synthesis rhythm (Harper et al. 2005). Several days following the therapy, the introduction of cells with an excessive amount of b-amyloid into the SCN of mice disrupted their activity and regular body temperature patterns (Tate et al. 1992). Research has shown that b-amyloid leads to an increase in the breakdown of Bmal1 protein, as stated by Musiek and Holtzman in 2016. The hypothesis suggests that the synchronisation between the main and peripheral oscillators is disturbed in Alzheimer's disease. This is believed to be caused by similar alterations in the expression of the CLOCK gene in the pineal glands of both Alzheimer's disease patients and rats with a damaged SCN (Wu et al. 2006).

Huntington's disease is a genetic disorder that occurs in individuals who have inherited the huntingtin allele with a large number of CAG repeats (more than 36). It is an autosomal-dominant condition, meaning that only one copy of the mutated gene is needed for the disease to develop. The symptoms of this condition consist of hyperkinesia, dystonia, cognitive and mental abnormalities, sleep-wake rhythm disturbance, and neuronal degeneration in the striatum and hypothalamus regions (Kassubek et al. 2004). Huntington's disease mouse models consist of R6/2 lines that include either 140 or 250 CAG repeats in the Htt gene, as well as transgenic BACHD knock-in mice that carry an artificial bacterial chromosome with numerous human Htt repeats. Both models demonstrate anomalies in rest-activity and body temperature circadian rhythms before the onset of neurological symptoms (Musiek and Holtzman 2016). Disruption of circadian rhythms (CR) and increased expression of clock genes (CGs) seem to be a secondary condition associated with damage to neuronal networks in both mice models and individuals with neurodegenerative illnesses. For example, the cultured SCN tissue from R6/2 mice exhibited typical CG expression, although the SCN of the same mice in vivo did not. Animal studies have shown that circadian disturbance in both humans and laboratory rats is linked to metabolic diseases and obesity (Lucassen et al. 2016; Wang et al. 2015). To ensure metabolic balance, the timing of metabolic gene expression must align with the timing of food intake. During the circadian cycle, the expression of CGs in b-cells, the levels of hormones that control nutritional behaviour and glucose metabolism,

Table 4.2 Pattern of clock gene expression in rodent models of diabetes mellitus

Model	Tissue	Effect on clock gene expression	Other effects
Rats, streptozotocin administration	Heart	Phase shift of Bmal1, Clock, Cry1, 2, Per1, Per2, Per3 gene expressions	Increase in serum glucose levels, blood glucose rhythm fluctuations
Mice, streptozotocin administration	Lung, GI tract, kidney	Phase shift of Per2, Per3 expression	Increase in serum glucose levels and no changes in the feeding rhythms
Administration of streptozotocin in neonatal mice	Liver	Increase in Bmal1, Cry1, Per 1 gene expression	Increase in serum glucose levels, glucose metabolism gene expression disruption
kk-Ay diabetic mice	Liver	Per3, Bmal1, and Npas2 gene expression rhythm disruption	Increase in serum glucose levels, changes in the rhythm of expression of 13 genes
Goto-kakizaki rats	Liver, fat tissue	No difference in the rhythm of the Clock, Bmal1, and Cry1	Increase in serum glucose levels
Diabetic Zucker rats with obesity	SCN, liver Aorta, heart, fat	Changes in Cry1 and Per1 gene expression	Increase in serum glucose levels

and the mRNA of key enzymes that determine the rates of glycolysis, oxidative phosphorylation, and the oxidation of fatty acids all vary (Marcheva et al. 2009; Vieira et al. 2014).

Metabolic diseases have an impact on the expression of CR and CGs in several diabetic models produced in mice, either genetically or chemically (Table 4.2). CG mutations have a detrimental impact on circadian metabolic organisation, leading to increased susceptibility to metabolic syndrome in knockout/mutant rats.

4.4.2 Other Rodent Model Systems of Circadian Clock Disruption

Circadian rhythm disruption refers to the condition where the cycles of rest and activity, as well as other internal rhythms, are not aligned with the natural cycles of light and darkness in the environment, and are not regulated accordingly. Rhythm disruption, typically linked to the ageing process, is also common among adults due to factors such as artificial light pollution, shift work, travel across several time zones, and the frequent use of electronic devices and nighttime illumination. Furthermore, the circadian system is compromised by the variable length of day and night, as stated by Anisimov et al. in 2012. Multiple research investigating shift and night work has revealed the detrimental impact it has on human health. These impacts include disruptions in sleep patterns, mood disorders, impaired metabolism and obesity, abnormalities in cardiovascular function, gastrointestinal issues, and an elevated chance of developing cancer (Touitou et al. 2016). While there is a lack of

Table 4.3 Effect of light regime on ageing and tumour development in different mice strains. Note the high-rate negative effects of clock-disrupting light schedules

Strain	Light schedules	Mean lifespan	% of tumour-bearing mice	Total number of tumours
HER-2/neu transgenic mice	Light: dark	281 ± 8.1 days	76.7	75
	Light: light	308 ± 10.8 days	76	103
129/Sv	Light: dark	763 ± 21.8 days	86.7	39
	Light: light	701 ± 20.7 days	81.4	38

methodologically rigorous research on the effects of caloric restriction (CR) on ageing in humans, studies on animals have shown that disrupting the circadian rhythm can lead to accelerated ageing. These studies evaluate markers of ageing and longevity.

Remarkably, mice that were exposed to a long-term jet lag regimen, resembling the rotating-shift schedule of nurses, displayed a comparable decrease in the amplitude of their activity circadian rhythm, despite a 12-h disparity in the peak of their activity (Rea et al. 2008). Various strategies are employed to disturb the circadian clock in laboratory animals, such as manipulating the light mode, performing surgical procedures, and creating molecular clock mutations or knockout models.

4.4.3 Light/Photoperiod

The experimental scenarios are either kept constant with either continuous darkness (LL) or continuous light (DD), or they are altered to simulate irregular light cycles, such as those experienced during jet lag or in regions with natural light variations. Dyslexia or visual impairment disrupts the regularity of rhythm, yet internal coordination may still be maintained due to these problems. The melatonin and SCN output signals remain intact. Continuous illumination, inversion of the light-dark rhythm, and surgical removal of the pineal gland markedly decrease the synthesis of melatonin. A short period of exposure to LL is enough to remove or greatly postpone the natural cycles of prolactin and corticosterone in mice (Claustrat et al. 2008). Low light circumstances also modify the expression of clock genes in the SCN (Sudo et al. 2003). As a result of the degeneration of neural circuits, the suprachiasmatic nucleus (SCN) in aged rats becomes more vulnerable to the desynchronizing effect of constant light (LL) exposure (Polidarová et al. 2017). Chronic disruption of the circadian rhythm caused by exposure to light conditions or surgical procedures leads to a decrease in the average lifespan of rodents, hampers glucose metabolism, triggers a phenotype similar to metabolic syndrome, affects the immune response, reduces physical activity and performance, results in cognitive decline, and increases the occurrence of spontaneous tumour formation, among other effects. Nonetheless, the impact of the light schedule on the process of ageing and the development of cancer is highly reliant on the genetic makeup of the mouse strain, as indicated in Table 4.3.

4.5 Effects of Mutations in Circadian Genes on Circadian Rhythms and Age-Related Disorders

Period gene mutation was the first circadian mutation discovered in Drosophila. However, the circadian behaviour of such mutant rodents is comparable to that of their wild-type littermates. Each family of duplicated clock genes in mammals contains two or more members. These genes are at least partially redundant, with the exception of Bmal2, which cannot be activated in the absence of Bmal1. When one of the paralogs is absent for other CGs, the other may be activated. Therefore, Clock knockout rodents with a complete absence of this gene express Npas2 in their tissues and exhibit a nearly normal phenotype. Simultaneously, rodents with a mutant Clock express a nonfunctional protein and display other pathological phenotypic characteristics. Nevertheless, the maintenance of both functional Per and Cry family members is essential for intrinsic rhythmicity strength and entrainment range. Erzberger and colleagues (2013) found that mutant mice have a reduced amplitude of behavioural rhythm, a shorter period, and are better able to entrain to short 16 h LD cycles. Cry1/Cry2 double-mutant rats have abnormal wheel-running behaviour and a disruption in the rhythmic expression of clock genes when kept in constant darkness. However, exposure to light partially recovers their activity rhythm (Albus et al. 2005).

Mice that do not possess CGs establish a correlation between disturbances in the circadian rhythm, the process of ageing, and the development of age-related diseases. For example, animals who have had their Clock and Bmal1 genes removed experience a reduced lifespan and significant changes in their metabolism. Per2/ animals are prone to developing cancer, and in certain cases, females undergo reproductive ageing at an early stage. Knocking down tissue-specific clock genes allows scientists to uncover the function of peripheral oscillators in controlling various physiological processes and eliminate the impact of the central clock. Furthermore, researchers have developed models that allow for the targeted activation of a single gene in a global knockout organism using a specific promoter. This enables the expression of the gene of interest only in tissues where the promoter is active.

However, non-circadian effects of CGs during prenatal and neonatal development may modulate the phenotypic characteristics of knockout rodents. In the case of Bmal/rodents, inducible knockouts that express this gene until puberty do not exhibit a shortened lifespan or arthropathies. Still, their behaviour becomes erratic when the gene is turned off (Yang et al. 2016).

4.6 Optimising Circadian Rhythms and Extending Longevity in Rodents Via Interventions

The proper functioning of the circadian system is essential for health maintenance and longevity. Wyse and colleagues (2010) found a correlation between the proximity of the free-running period tau to 24 h and the lifespan of laboratory mouse strains. As aging itself causes a decline in circadian output, circadian disruption or age-related diseases may exacerbate this decline significantly. Light is the most influential zeitgeber for nocturnal and diurnal species alike. It has been demonstrated that SCN neuronal activity is firmly correlated with photic input (Nakamura et al. 2008). Bright light therapy is a promising method for treating neurological conditions (Schroeder and Colwell 2013). Nonetheless, the circadian system may respond to additional entrainment cues. Some of the experimental data pertaining to environmental signals (light therapy) and various chronobiotics treatments are discussed below.

Stress and physical activity affect the hypothalamic–pituitary–adrenal and sympathetic–adrenal–medullary axis by altering the synthesis of adrenal hormones (Mastorakos and Pavlatou 2005). In the absence of other entraining stimuli, induced exercise altered rest-activity cycles in Syrian hamsters confined to DD (Reebs and Mrosovsky 1989). The timing of exercise or stressing stimuli is important: phase advance is observed before the subjective middle of the day, whereas phase-delaying effects are observed at the conclusion of the activity phase. It has been shown that scheduled activity or stress in either the first or second half of active phase under LD or DD conditions shifts behavioural rhythms and CGs expression differentially (Reebs and Mrosovsky 1989; Schroeder and Colwell 2013).

Participation in regular physical activity can be beneficial in repairing cognitive reserves that have been negatively affected by the natural process of ageing or genetic factors. Leise et al. (2013) discovered that in older mice, the implementation of a prescribed wheel running regimen enhanced behavioural rhythmicity compared to mice that did not have access to a wheel. A study (Schroeder and Colwell 2013) found that regular exercise has positive effects on the time and quality of sleep in older adults. Despite the strong influence of physical exercise on the body's internal clock, the specific ways in which physical activity affects the circadian system at a molecular level are still not understood.

Clearly, nourishment takes place during the active period, with digestive and metabolic processes happening before it. When food intake is limited to specific time periods, the peripheral oscillators (such as the liver, heart, kidney, pancreas, adipose tissue, and gastrointestinal tract) synchronise with the meal schedule rather than the schedule of light exposure. SCN-lesioned mice, who lack central pacemaker regulation, exhibit faster entrainment to food cues compared to control mice, as observed in a study by Potter et al. in 2016. When peripheral oscillators become unsynchronized under low light (LL) circumstances, timed feeding acts as a stimulus that aligns their rhythms (Hamaguchi et al. 2015).

The SCN exhibits a bidirectional neuronal link with sleep centres. The circadian system governs sleep patterns and receives input from other brain areas on the status of being asleep or awake (Deboer et al. 2003). Pharmacologically induced sleep in a mouse HD model enhanced cognitive and motor functioning and rhythmic Per2 expression (Pallier et al. 2007). Therefore, ensuring a sufficient amount of sleep or using medication to induce regular sleep could potentially improve the rhythm and overall quality of life in older individuals with neurodegenerative illnesses.

Due to the interconnectedness of most biological signalling networks, many medicines have numerous targets. The phenomenon known as the six degrees of separation seems to be applicable to CGs at the molecular level. Certain well-established medications have the ability to alter or restore the molecular clock. This can have both positive and negative effects. For example, it may be beneficial in cases of jet lag but detrimental for patients with phase advance syndrome taking drugs that reduce the length of the circadian period. The substances that have an impact on circadian rhythms include caffeine, sildenafil, lithium, valproic acid, and serotonergic medicines (Schroeder and Colwell 2013). Nevertheless, each one has a clearly defined purpose, and certain ones can be habit-forming.

The production of melatonin is reduced at its peak in elderly rats and humans, which is thought to be a factor in the decrease of rhythmic functions associated with ageing. Moreover, this hormone deficit is involved in the development of several age-related illnesses. Exogenous melatonin can be used to improve sleep quality, regulate biological rhythms, and alleviate cardiovascular, metabolic, and neurological diseases. Studies have shown that melatonin can improve the synchronisation of biological rhythms, cognitive and motor abilities, the severity of metabolic disorders, and the occurrence of tumours in older mice (Anisimov et al. 2012; Schroeder and Colwell 2013). Melatonin's antioxidant and anticancer characteristics make it potentially vital for promoting healthy ageing. In a study conducted by Morley and colleagues (2012), it was found that melatonin has the ability to decrease inflammation, oxidative stress, neuronal loss, and mitochondrial damage in the brains of SAMP8 mice. Additionally, melatonin was found to enhance cognitive function. Furthermore, some research has shown evidence for the anti-aging effects of melatonin (Claustrat et al. 2008; Anisimov et al. 2012). The impact of melatonin on lifetime extension remains uncertain, as it is not obvious whether this is primarily attributed to its chronobiological or antioxidant properties or other intricate mechanisms.

Ultimately, the environmental disruptions to the circadian system, known as CRs, that arise with age or are associated with age-related illnesses can be rectified through the implementation of scheduled physical activity, regulated nutrition, exposure to moderate levels of stress, or the administration of chronobiotics medication.

4.7 Conclusion

Multiple cellular oscillators in the mammalian circadian system are entrained and synchronized with each other and external time by the central pacemaker in the SCN, which receives non-visual photic stimuli from the retina. In order for the SCN to synchronize peripheral rhythms, neural signals and hormone levels, particularly melatonin, must be regulated.

Endogenous circadian disturbance is a result of the process of ageing. The circadian system experiences a decrease in sensitivity to external stimuli and a decrease in its ability to adjust (entrain) to changes in external rhythms. Additionally, the amplitude and rhythms of the circadian system become fragmented. Anatomical diminishment is noted in the innervation of retinal ganglion cells and the pineal gland. SCN neurons experience a loss of synchronicity in both their firing patterns and the production of clock genes. The age-related disruption of circadian gene expression occurs in specific tissues.

The association between ageing and circadian disruption can be explained by several factors. Firstly, certain strains of senescence-accelerated mice show disrupted circadian rhythms. Secondly, animals with mutations in CG genes display early age-related alterations. Lastly, circadian disruption has been found to enhance both the frequency and severity of age-related disorders.

Although their circadian system is less sensitive and entraining stimuli must be more potent than normal physiological signals, certain interventions may be able to rescue CR in elderly animals. Certain pharmacological agents also have an effect on both CR and senescence.

References

Albus H, Vansteensel MJ, Michel S, Block GD, Meijer JH (2005) A GABAergic mechanism is necessary for coupling dissociable ventral and dorsal regional oscillators within the circadian clock. Curr Biol 15(10):886–893

Anisimov VN, Vinogradova IA, Panchenko AV, Popovich IG, Zabezhinski MA (2012) Light-at-night-induced circadian disruption, cancer and aging. Curr Aging Sci 5(3):170–177

Aujard F, Herzog E, Block G (2001) Circadian rhythms in firing rate of individual suprachiasmatic nucleus neurons from adult and middle-aged mice. Neuroscience 106(2):255–261

Bell-Pedersen D, Cassone VM, Earnest DJ, Golden SS, Hardin PE, Thomas TL, Zoran MJ (2005) Circadian rhythms from multiple oscillators: lessons from diverse organisms. Nat Rev Genet 6(7):544–556. https://doi.org/10.1038/nrg1633

Boulamery A, Simon N, Vidal J, Bruguerolle B (2010) Effects of L-Dopa on circadian rhythms of 6-OHDA striatal lesioned rats: a radiotelemetric study. Chronobiol Int 27(2):251–264

Brock MA (1991) Chronobiology and aging. J Am Geriatr Soc 39(1):74–91. https://doi.org/10.1111/j.1532-5415.1991.tb05909.x

Buysse DJ (2005) Diagnosis and assessment of sleep and circadian rhythm disorders. J Psychiatr Pract 11(2):102–115

Cernuda-Cernuda R, Huerta JJ, Muñoz Llamosas M, Alvarez-Uría M, García-Fernández JM (2000) Age-related morphometric changes in the pineal gland. A comparative study between C57BL/6J and CBA mice. Histol Histopathol 15(4):1087–1092. https://doi.org/10.14670/HH-15.1087

Chang H-C, Guarente L (2013) SIRT1 mediates central circadian control in the SCN by a mechanism that decays with aging. Cell 153(7):1448–1460

Claustrat B, Valatx J-L, Harthe C, Brun J (2008) Effect of constant light on prolactin and corticosterone rhythms evaluated using a noninvasive urine sampling protocol in the rat. Horm Metab Res 40(6):398–403

Davidson AJ, Straume M, Block GD, Menaker M (2006) Daily timed meals dissociate circadian rhythms in hepatoma and healthy host liver. Int J Cancer 118(7):1623–1627

Deboer T, Vansteensel MJ, Détári L, Meijer JH (2003) Sleep states alter activity of suprachiasmatic nucleus neurons. Nat Neurosci 6(10):1086–1090

Dijk DJ, Beersma D, Daan S, Lewy AJ (1989) Bright morning light advances the human circadian system without affecting NREM sleep homeostasis. Am J Phys Regul Integr Comp Phys 256(1):R106–R111

Dubocovich ML, Delagrange P, Krause DN, Sugden D, Cardinali DP, Olcese J (2010) International Union of Basic and Clinical Pharmacology. LXXV. Nomenclature, classification, and pharmacology of G protein-coupled melatonin receptors. Pharmacol Rev 62(3):343–380. https://doi.org/10.1124/pr.110.002832

Duffy JF, Zitting K-M, Chinoy ED (2015) Aging and circadian rhythms. Sleep Med Clin 10(4):423–434

Duncan MJ, Smith JT, Franklin KM, Beckett TL, Murphy MP, Clair DKS, Donohue KD, Striz M, O'Hara BF (2012) Effects of aging and genotype on circadian rhythms, sleep, and clock gene expression in APPxPS1 knock-in mice, a model for Alzheimer's disease. Exp Neurol 236(2):249–258

Erzberger A, Hampp G, Granada A, Albrecht U, Herzel H (2013) Genetic redundancy strengthens the circadian clock leading to a narrow entrainment range. J R Soc Interface 10(84):20130221

Farajnia S, Meijer JH, Michel S (2015) Age-related changes in large-conductance calcium-activated potassium channels in mammalian circadian clock neurons. Neurobiol Aging 36(6):2176–2183

Grandner MA, Martin JL, Patel NP, Jackson NJ, Gehrman PR, Pien G, Perlis ML, Xie D, Sha D, Weaver T (2012) Age and sleep disturbances among American men and women: data from the US Behavioral Risk Factor Surveillance System. Sleep 35(3):395–406

Hamaguchi Y, Tahara Y, Kuroda H, Haraguchi A, Shibata S (2015) Entrainment of mouse peripheral circadian clocks to <24 h feeding/fasting cycles under 24 h light/dark conditions. Sci Rep 5(1):14207

Harper DG, Volicer L, Stopa EG, McKee AC, Nitta M, Satlin A (2005) Disturbance of endogenous circadian rhythm in aging and Alzheimer disease. Am J Geriatr Psychiatry 13(5):359–368

Hastings M, Brancaccio M, Maywood E (2014) Circadian pacemaking in cells and circuits of the suprachiasmatic nucleus. J Neuroendocrinol 26(1):2–10

Hayashi Y, Koyanagi S, Kusunose N, Okada R, Wu Z, Tozaki-Saitoh H, Ukai K, Kohsaka S, Inoue K, Ohdo S (2013) The intrinsic microglial molecular clock controls synaptic strength via the circadian expression of cathepsin S. Sci Rep 3(1):2744

Hazlerigg DG, Barrett P, Hastings M, Morgan PJ (1996) Are nuclear receptors involved in pituitary responsiveness to melatonin? Mol Cell Endocrinol 123(1):53–59

Honma S, Ono D, Suzuki Y, Inagaki N, Yoshikawa T, Nakamura W, Honma K (2012) Suprachiasmatic nucleus: cellular clocks and networks. Prog Brain Res 199:129–141

Jud C, Schmutz I, Hampp G, Oster H, Albrecht U (2005) A guideline for analyzing circadian wheel-running behavior in rodents under different lighting conditions. Biol Proced Online 7(1):101–116. https://doi.org/10.1251/bpo109

Kassubek J, Juengling FD, Kioschies T, Henkel K, Karitzky J, Kramer B, Ecker D, Andrich J, Saft C, Kraus P, Aschoff AJ, Ludolph AC, Landwehrmeyer GB (2004) Topography of cerebral atrophy in early Huntington's disease: a voxel based morphometric MRI study. J Neurol Neurosurg Psychiatry 75(2):213–220

Kunz D, Schmitz S, Mahlberg R, Mohr A, Stöter C, Wolf K-J, Herrmann WM (1999) A new concept for melatonin deficit: on pineal calcification and melatonin excretion. Neuropsychopharmacology 21(6):765–772

Leise T, Harrington ME, Molyneux P, Song I, Queenan H, Zimmerman E, Lall G, Biello S (2013) Voluntary exercise can strengthen the circadian system in aged mice. Age 35(6):2137–2152

Li H, Satinoff E (1998) Fetal tissue containing the suprachiasmatic nucleus restores multiple circadian rhythms in old rats. Am J Phys Regul Integr Comp Phys 275(6):R1735–R1744. https://doi.org/10.1152/ajpregu.1998.275.6.R1735

Li M-D, Li C-M, Wang Z (2012) The role of circadian clocks in metabolic disease. Yale J Biol Med 85(3):387–401

Lucassen EA, Coomans CP, van Putten M, de Kreij SR, van Genugten JH, Sutorius RP, de Rooij KE, van der Velde M, Verhoeve SL, Smit JW (2016) Environmental 24-hr cycles are essential for health. Curr Biol 26(14):1843–1853

Marcheva B, Ramsey KM, Affinati A, Bass J (2009) Clock genes and metabolic disease. J Appl Physiol 107(5):1638–1646

Mastorakos G, Pavlatou M (2005) Exercise as a stress model and the interplay between the hypothalamus-pituitary-adrenal and the hypothalamus-pituitary-thyroid axes. Horm Metab Res 37(9):577–584

Mattam U, Jagota A (2015) Daily rhythms of serotonin metabolism and the expression of clock genes in suprachiasmatic nucleus of rotenone-induced Parkinson's disease male Wistar rat model and effect of melatonin administration. Biogerontology 16:109–123

Mendoza J, Graff C, Dardente H, Pevet P, Challet E (2005) Feeding cues alter clock gene oscillations and photic responses in the suprachiasmatic nuclei of mice exposed to a light/dark cycle. J Neurosci 25(6):1514–1522

Morley JE, Farr SA, Kumar VB, Armbrecht HJ (2012) The SAMP8 mouse: a model to develop therapeutic interventions for Alzheimer's disease. Curr Pharm Des 18(8):1123–1130

Musiek ES, Holtzman DM (2016) Mechanisms linking circadian clocks, sleep, and neurodegeneration. Science 354(6315):1004–1008. https://doi.org/10.1126/science.aah4968

Nakamura W, Yamazaki S, Nakamura TJ, Shirakawa T, Block GD, Takumi T (2008) In vivo monitoring of circadian timing in freely moving mice. Curr Biol 18(5):381–385. https://doi.org/10.1016/j.cub.2008.02.024

Nakamura TJ, Nakamura W, Yamazaki S, Kudo T, Cutler T, Colwell CS, Block GD (2011) Age-related decline in circadian output. J Neurosci 31(28):10201–10205

Nakamura TJ, Nakamura W, Tokuda IT, Ishikawa T, Kudo T, Colwell CS, Block GD (2015) Age-related changes in the circadian system unmasked by constant conditions. eNeuro 2(4):ENEURO.0064

Nebes RD, Buysse DJ, Halligan EM, Houck PR, Monk TH (2009) Self-reported sleep quality predicts poor cognitive performance in healthy older adults. J Gerontol B Psychol Sci Soc Sci 64(2):180–187

Pallier PN, Maywood ES, Zheng Z, Chesham JE, Inyushkin AN, Dyball R, Hastings MH, Morton AJ (2007) Pharmacological imposition of sleep slows cognitive decline and reverses dysregulation of circadian gene expression in a transgenic mouse model of Huntington's disease. J Neurosci 27(29):7869–7878

Polidarová L, Sládek M, Novosadová Z, Sumová A (2017) Aging does not compromise in vitro oscillation of the suprachiasmatic nuclei but makes it more vulnerable to constant light. Chronobiol Int 34(1):105–117

Potter GD, Skene DJ, Arendt J, Cade JE, Grant PJ, Hardie LJ (2016) Circadian rhythm and sleep disruption: causes, metabolic consequences, and countermeasures. Endocr Rev 37(6):584–608

Rea MS, Bierman A, Figueiro MG, Bullough JD (2008) A new approach to understanding the impact of circadian disruption on human health. J Circadian Rhythms 6(1):1–14

Reebs SG, Mrosovsky N (1989) Large phase-shifts of circadian rhythms caused by induced running in a re-entrainment paradigm: the role of pulse duration and light. J Comp Physiol A 165(6):819–825

Roozendaal B, Van Gool WA, Swaab DF, Hoogendijk JE, Mirmiran M (1987) Changes in vasopressin cells of the rat suprachiasmatic nucleus with aging. Brain Res 409(2):259–264

Satinoff E, Kent S, Li H, Megirian D, Tomkowiak J (1991) Circadian rhythms of body temperature and drinking and responses to thermal challenge in rats after PCPA. Pharmacol Biochem Behav 38(2):253–257

Schmidt RE, Dorsey DA, Parvin CA, Beaudet LN (2006) Sympathetic neuroaxonal dystrophy in the aged rat pineal gland. Neurobiol Aging 27(10):1514–1523. https://doi.org/10.1016/j.neurobiolaging.2005.08.005

Schroeder AM, Colwell CS (2013) How to fix a broken clock. Trends Pharmacol Sci 34(11):605–619

Semo M, Peirson S, Lupi D, Lucas RJ, Jeffery G, Foster RG (2003) Melanopsin retinal ganglion cells and the maintenance of circadian and pupillary responses to light in aged rodless/coneless (rd/rd cl) mice. Eur J Neurosci 17(9):1793–1801

Silver R, LeSauter J, Tresco PA, Lehman MN (1996) A diffusible coupling signal from the transplanted suprachiasmatic nucleus controlling circadian locomotor rhythms. Nature 382(6594):810–813. https://doi.org/10.1038/382810a0

Sudo M, Sasahara K, Moriya T, Akiyama M, Hamada T, Shibata S (2003) Constant light housing attenuates circadian rhythms of mPer2 mRNA and mPER2 protein expression in the suprachiasmatic nucleus of mice. Neuroscience 121(2):493–499

Sujino M, Masumoto K, Yamaguchi S, van der Horst GT, Okamura H, Inouye S-IT (2003) Suprachiasmatic nucleus grafts restore circadian behavioral rhythms of genetically arrhythmic mice. Curr Biol 13(8):664–668

Tate B, Aboody-Guterman KS, Morris AM, Walcott EC, Majocha RE, Marotta CA (1992) Disruption of circadian regulation by brain grafts that overexpress Alzheimer beta/A4 amyloid. Proc Natl Acad Sci 89(15):7090–7094

Touitou Y, Touitou D, Reinberg A (2016) Disruption of adolescents' circadian clock: the vicious circle of media use, exposure to light at night, sleep loss and risk behaviors. J Physiol Paris 110(4):467–479

Turek FW, Dugovic C, Zee PC (2001) Current understanding of the circadian clock and the clinical implications for neurological disorders. Arch Neurol 58(11):1781–1787

Valentinuzzi VS, Scarbrough K, Takahashi JS, Turek FW (1997) Effects of aging on the circadian rhythm of wheel-running activity in C57BL/6 mice. Am J Phys Regul Integr Comp Phys 273(6):R1957–R1964

Van Gool WA, Mirmiran M (1986) Aging and circadian rhythms. Prog Brain Res 70:255–277. https://doi.org/10.1016/s0079-6123(08)64309-5

Videnovic A, Willis GL (2016) Circadian system—a novel diagnostic and therapeutic target in Parkinson's disease? Mov Disord 31(3):260–269

Vieira E, Burris TP, Quesada I (2014) Clock genes, pancreatic function, and diabetes. Trends Mol Med 20(12):685–693

Wang D, Chen S, Liu M, Liu C (2015) Maternal obesity disrupts circadian rhythms of clock and metabolic genes in the offspring heart and liver. Chronobiol Int 32(5):615–626

Weinert D (2000) Age-dependent changes of the circadian system. Chronobiol Int 17(3):261–283

Wu Y, Fischer DF, Kalsbeek A, Garidou-Boof M, van der Vliet J, van Heijningen C, Liu R, Zhou J, Swaab DF, Wu Y (2006) Pineal clock gene oscillation is disturbed in Alzheimer's disease, due to functional disconnection from the "master clock". FASEB J 20(11):1874–1876

Wyse C, Coogan A, Selman C, Hazlerigg D, Speakman J (2010) Association between mammalian lifespan and circadian free-running period: the circadian resonance hypothesis revisited. Biol Lett 6(5):696–698

Yang G, Chen L, Grant GR, Paschos G, Song W-L, Musiek ES, Lee V, McLoughlin SC, Grosser T, Cotsarelis G (2016) Timing of expression of the core clock gene Bmal1 influences its effects on aging and survival. Sci Transl Med 8(324):324ra16

Zhang J, Dong X, Fujimoto Y, Okamura H (2004) Molecular signals of mammalian circadian clock. Kobe J Med Sci 50(3/4):101

Circadian Control of Mitochondrial Dynamics and Its Implication in Aging

5

5.1 Introduction

Mitochondria are essential for the generation of energy through oxidative metabolism. Their malfunction has been associated with metabolic disorders and ageing (Chan 2006); however, a definitive cause-and-effect relationship has not always been proven. Aside from dietary availability, the synthesis of ATP is also regulated by various cellular mechanisms that are tailored to fulfil certain metabolic needs. Uncoupled respiration in brown adipocytes, which generates heat by releasing oxidative energy, is an extraordinary illustration (Wikstrom et al. 2014). Mitochondrial adaptation to different nutritional situations is influenced by structural changes regulated by mitochondrial dynamics, which involve mitochondrial fusion, fission, and selective autophagy (mitophagy). The molecular constituents and functions within cells of mitochondrial dynamics were first elucidated through the utilisation of *Saccharomyces cerevisiae* and model organisms (Westermann 2010). Egner and colleagues (2002) showed that the shape of mitochondria can change in response to exposure to various carbon sources or growth conditions. The process of mitochondrial dynamics is essential for regulating both the generation of energy and the maintenance of quality control within the cell (Liesa and Shirihai 2013; Youle and Van Der Bliek 2012). The significance of maintaining mitochondrial quality control is demonstrated by various human disorders, such as age-related neurodegeneration. Research has shown that mutations in genes involved in mitophagy are linked to autosomal recessive Parkinson's disease (Valente et al. 2004).

This chapter will address the role of mitochondrial dynamics in energy metabolism. The regular pattern of eating and refraining from eating leads to changes in the amount of nutrients, energy needed, and the level of oxidative stress. In metabolically active organs like the liver, there is a delicate balance between bioenergetic adaptability, which refers to the ability to adjust energy production, the formation of reactive oxygen species (ROS) as a result of oxidative metabolism, and the removal of damaged parts of mitochondria. First, we will analyse the evidence that supports

S. I. Rizvi, G. Majumdar, *Circadian Clock and Aging*,
https://doi.org/10.1007/978-981-97-6695-6_5

the involvement of mitochondrial dynamics in this process. We will examine how the circadian clock controls mitochondrial dynamics and oxidative metabolism and explore the importance of this regulation in the process of ageing and age-related disorders, as the circadian rhythm is a crucial part of the feeding-fasting cycle.

5.2 Cellular Function Involved in Mitochondrial Dynamics

5.2.1 The Machinery

Mitochondria can modify their morphology in response to different nutritional situations, thanks to the machinery of mitochondrial dynamics. Under conditions of low and high food availability, mitochondria in cultivated cells exhibit elongated and fragmented networks, respectively (Liesa and Shirihai 2013). This adaptation performs various crucial activities, such as optimising metabolic efficiency, regulating oxidative stress, and ensuring quality control of mitochondria (see to Fig. 5.1 for further information on mitochondrial quality control). Mitochondria have a lipid membrane consisting of two layers. The matrix is enclosed by the inner membrane, which is folded into cristae. These cristae include membrane-bound enzymes for the electron transport chain and ATP-synthase. Mitochondrial fission is the process of dividing mitochondria into two distinct entities, each with its own matrix and intermembrane space. The process of fission is facilitated by the activation of a cytosolic adaptor called dynamin-related protein 1 (DRP1). DRP1 binds to receptors on the outer membranes of mitochondria, such as mitochondrial fission factor-MFF, mitochondrial fission 1 protein-FIS1, MiD49, and MiD51. This binding causes the mitochondria to be connected and initiates their separation by cutting both the inner and

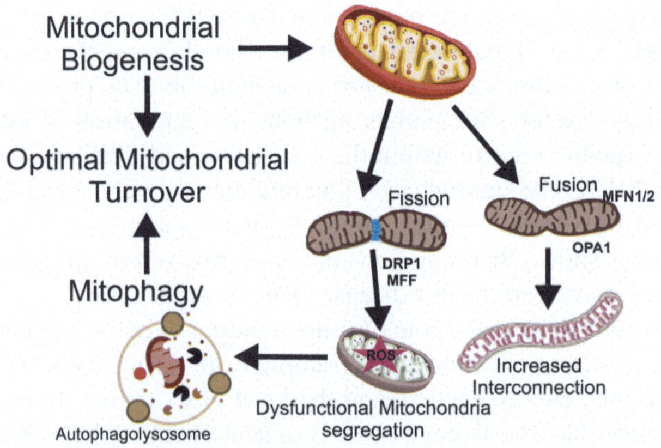

Fig. 5.1 Quality control pathways within mitochondria preserve mitochondrial function. Optic atrophy protein 1 (OPA1) and mitofusins 1 and 2 (MFN1/2) facilitate mitochondrial fusion, which generates a network of interconnected, healthy mitochondria. Dynamin-related protein 1 (DRP1) and mitochondrial fission factor (MFF) proteins are essential for mitochondrial fission. By means of mitophagy, fragmented, dysfunctional mitochondria are eliminated

outer membranes (Chang and Blackstone 2010; Losón et al. 2013). Fusion is a biphasic mechanism that connects mitochondria using membrane-embedded GTPases. Mitofusin 1 and 2 (MFN1 and 2) are responsible for merging the outer membranes to combine the intermembrane gaps (Rojo et al. 2002; Santel and Fuller 2001). Alexander et al. (2000) discovered that OPA1 facilitates the fusion of the inner membranes, which leads to the fusion of the matrices. Phosphorylation of DRP1 at serine 637 can accelerate fusion by suppressing fission, as demonstrated by Gomes et al. in 2011. The bulk of mitochondria is sustained through an ongoing process of biosynthesis and decay.

Mitophagy is a specific type of autophagy that focuses on removing and breaking down damaged mitochondria in a selective manner. The Ser/Thr kinase PINK1 and the E3 ubiquitin ligase PARKIN work together in this process of disposal. When the mitochondria lose their electrical potential, PINK1 avoids being broken down by the protease PARL, which is constantly breaking down proteins in the inner membrane, and instead builds up on the outside membrane. Subsequently, it phosphorylates PARKIN to fully activate its ubiquitin ligase activity (Chen and Dorn 2013; Koyano et al. 2014). PARKIN that has been activated is recruited to mitochondria and adds ubiquitin molecules to outer membrane proteins. This leads to the engulfment of mitochondria in isolation membranes, which then combine with lysosomes. Youle and Narendra (2011) propose that the ubiquitin-binding adaptor P62 potentially contributes to the initiation of autophagosome formation through the involvement of PARKIN. Furthermore, the BCL2/adenovirus E1B interacting protein 3 (BNIP3) has the ability to directly interact with LC3-II through its N-terminal region. The lipidated form of LC3 enables the fusion of autophagosomes with lysosomes, resulting in the formation of autolysosomes, where mitochondria are broken down.

5.2.2 Mitochondrial Dynamics: Biological Sensor of Bioenergetics and Oxidative Stress

Mitchell (1961) defines oxidative phosphorylation as the coupling of substrate oxidation by the electron transport chain to the generation of a trans-membrane proton gradient (Mitchell 1961). Subsequently, the protons are transported back to the matrix through the ATP synthase proton pore, leading to the synthesis of ATP.

Nobes and colleagues (1990) found that the proton leak in hepatocytes, mostly caused by mitochondrial anion carrier proteins, contributes to around 40% of the hepatocytes' resting metabolic rate. Electron leakage occurs when electrons exit the electron transport chain and react with oxygen to produce reactive oxygen species, specifically superoxide (O_2). This leads to the electron loss. Furthermore, there is a mutual dependence between proton and electron leaks (Jastroch et al. 2012), and a higher proton leak can lead to a reduced production of superoxide. The redox state seems to have the ability to sense and regulate mitochondrial dynamics. Highly nutritious conditions necessitate a significant emphasis on managing oxidative damage. Significantly, fission increases the level of uncoupled respiration, which, consequently, may decrease the quantity of superoxide generated (Liesa and Shirihai 2013). In contrast, mitochondrial fusion and elongation take place during periods of low nutritional conditions, when cells rely heavily on ATP production and require increased ATP generation efficiency (Gomes et al. 2011). This is due to the fact that cells in these environments necessitate a greater degree of ATP production efficiency. The physiological function of fusion and fission in mitochondria relies on the specific circumstances in which it takes place. Brown adipose tissue provides an extreme illustration of energy diversion towards thermogenesis instead of ATP generation by fission, regardless of calorie situation. This process is regulated by DRP1 via adrenergic and protein kinase A (PKA) signalling (Wikstrom et al. 2014). As mentioned earlier, this happens irrespective of the energy state of the tissue.

5.2.3 Quality Control and the Mitochondrial Dynamics Involved

The mitochondrial proteome, which consists of around 1500 proteins, is mostly encoded by the nucleus, accounting for around 99%. Therefore, maintaining proper protein balance, known as proteostasis, is crucial for the overall health of mitochondria. When used to suppress mitochondrial translation in mammalian cells, doxycycline causes an imbalance in the ratio of proteins between the mitochondria and the nucleus. This imbalance leads to a decrease in mitochondrial respiration and encourages the process of mitochondrial fission (Moullan et al. 2015). Excessive or uncontrolled generation of ROS can cause damage to mitochondrial proteins, lipids, and DNA. Damage to the electron transport chain might exacerbate the generation of reactive oxygen species (ROS). Mitochondria utilise many quality-control mechanisms, such as anti-oxidation enzymes, mitochondrial chaperones and proteases, and a mitochondria-associated degradation (MAD) route for outer-membrane proteins, to counteract these risks (Tanaka et al. 2010). Mitophagy is different from these mechanisms since it specifically breaks out faulty mitochondria. Mitophagy is linked to maintaining mitochondrial quality control during times of stress, such as when there are changes in mitochondrial proteostasis or increased oxidative stress. When a misfolded protein in the mitochondria is overproduced in Drosophila, it leads to a mitochondrial phenotype similar to the phenotype observed in Pink1 or Parkin mutants. However, this phenotype may be reversed by overproducing Parkin protein (Pimenta De Castro et al. 2012). Clark and his colleagues discovered that the absence of Drosophila PINK1 enhances susceptibility to many forms of damage, including oxidative stress (Clark et al. 2006). Furthermore, the liver mitochondria of Bnip3-deficient animals exhibit a decline in membrane potential, an abnormal shape, and a reduction in oxygen consumption, despite an increase in bulk (Glick et al. 2012). Nevertheless, it is worth noting that every cell comprises numerous mitochondria, ranging from tens to hundreds, each with its own circular double-stranded DNA genome (mtDNA) enclosed within nucleoids (Schon and Gilkerson 2010). Heteroplasmy refers to the presence of many variations of mutant mitochondrial DNA (mtDNA) alongside the normal, wild-type version within a single cell. Complementation has been shown to occur between wild-type (WT) and mutant mitochondria in rodents with mutant mitochondrial DNA (mtDNA) that has a loss of 5000 bp (Nakada et al. 2001). The presence of cytochrome *c* oxidase in individual mitochondria inside cells with mutant and wild-type mtDNA indicates content mixing. This supplementation would offset the deficiencies caused by mutations in mitochondrial DNA. Given that nucleoids do not exchange DNA (Schon and Gilkerson 2010), heteroplasmic mitochondria compensate for this by exchanging RNA and/or proteins to complement each other. Chen and colleagues subsequently showed that mitochondrial fusion is the mechanism that allows for complementation. It is worth mentioning that animals lacking Mfn1 and Mfn2 in their skeletal muscles experience a significant decline in muscle function and a decrease in mitochondrial DNA levels (Chen et al. 2010).

5.3 Circadian Rhythms and Regulation of Mitochondrial Activity

5.3.1 Metabolic Rhythms of Metabolism

It is crucial for mitochondria to adjust to variations in food availability and energy utilisation caused by alternating periods of feeding and fasting is a substantial challenge. The circadian clock has been associated with controlling many metabolic processes related to feeding and fasting responses, especially in research involving diurnal and nocturnal rats (Bass and Takahashi 2010). Recent studies have demonstrated that the circadian clock also has a role in regulating mitochondrial dynamics and oxidative metabolism (Jacobi et al. 2017; Peek et al. 2013). This regulatory mechanism ensures metabolic flexibility by maintaining a robust mitochondrial population and regulating oxidative stress (Fig. 5.2). Neufeld-Cohen and colleagues discovered that a significant fraction of proteins in the mitochondria of mice undergo

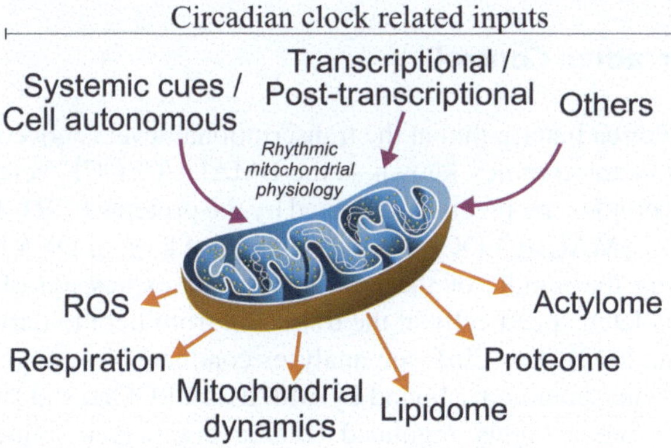

Fig. 5.2 Schematic representation of crucial input mechanisms that regulate a variety of circadian mitochondrial outputs. Reactive oxygen species (ROS). The purple arrows represent the input factors, the orange arrows are the outputs

diurnal oscillations. These proteins include those involved in the respiratory chain as well as enzymes responsible for lipid and glucose metabolism (Neufeld-Cohen et al. 2016). Several transcripts associated with mitochondria maintain a consistent level but are translated in a rhythmic manner (Atger et al. 2015). This rhythmic translation may be affected by changes in the length of poly(A) tails, which show diurnal fluctuations in the liver of mice (Kojima et al. 2012).

Post-translational changes, particularly acetylation, regulate mitochondrial function in the liver of mice (Hebert et al. 2013; Zhao et al. 2010). Mice who do not have the mitochondrial NAD+-dependent deacetylase sirtuin-3 (SIRT3) exhibit changes in the acetylation of mitochondrial proteins and experience a decline in mitochondrial function. The circadian clock regulators Bmal1/Clock influence mitochondrial activities by controlling the transcription of the rate-limiting enzyme involved in NAD+ production, known as nicotinamide phosphoribosyltransferase (NAMPT) (Peek et al. 2013; Ramsey et al. 2009). NAD+ oscillations influence SIRT3 activity, which in turn controls the acetylation of mitochondrial proteins (Peek et al. 2013). The diurnal variations in fatty acid oxidation and the synchronised changes in NAD+ levels happen concurrently. Hepatocytes that do not have Bmal1 show a decrease in the process of breaking down fatty acids and an increase in the excessive acetylation of SIRT3 targets. Peek et al. (2013) have shown that the injection of nicotinamide mononucleotide (NMN), a product of NAMPT, can restore the acetylation profiles of SIRT3 targets and improve mitochondrial oxidative capacity. Thus, it may be deduced that the dysfunction of mitochondria in liver cells lacking Bmal1 depends on the reduction in NAD+ levels. Additionally, there is a proposition suggesting that the cyclical presence of NAD+ can impact the functioning of SIRT1. This protein regulates the metabolic alterations observed during the transition from fasting to feeding. The regulation is accomplished by modulating various transcription factors, including Peroxisome Proliferator-Activated Receptor Gamma Coactivator 1-alpha (PGC1a), which is responsible for controlling the formation of mitochondria (Marcheva et al. 2013).

5.3.2 The Circadian Control

The regulation of circadian rhythm at the transcriptional level is governed by a central transcription factor complex known as the BMAL1/CLOCK heterodimer. The activities of this complex are precisely adjusted by the proteins PER1, PER2, PER3, CRY1, and CRY2. BMAL1/CLOCK reaches its highest level of DNA binding in the mouse liver during the middle of the light cycle. The expression of target genes peaks a few hours later, specifically at the transition from light to dark (Gong et al. 2015; Koike et al. 2012). The ChIP-seq analyses conducted by Koike et al. (2012) suggest that the genes commonly bound by BMAL1, CLOCK, and CRY1 indicate that mitochondria are a highly regulated cellular component influenced by the molecular clock. This implies that the circadian clock has the ability to anticipate the upcoming dark/feeding cycle by enhancing mitochondrial function, potentially in response to increased nutrient influx and metabolic stress. The circadian clock is expected to directly affect gene pathways related to mitochondrial fission (Fis1 and Mtfr1) and mitophagy (Pink1 and Bnip3). These pathways, along with the previously known activities of Bmal1 in oxidative metabolism and fatty acid oxidation, are likely to be influenced by the circadian clock. As observed using electron microscopy and Cox8-GFP tagging, the liver's mitochondrial morphology and network exhibit a diurnal pattern. During the light cycle, the networks appear more extended, whereas during the dark cycle, they appear more fragmented. Unlike the mitochondrial network of cultured cells, which quickly adjusts to different dietary conditions, this diurnal pattern remains unchanged even when the food supply is temporarily moved to the daylight. The hepatic mitochondria of mice with a liver-specific Bmal1 deletion (LBmal1KO) exhibit increased size consistently throughout the day, reduced respiratory activity, and heightened production of reactive oxygen species. Similarly, primary hepatocytes derived from LBmal1KO mice lack the ability to alter their shape and adapt their mitochondrial respiration when grown in nutrient-poor or nutrient-rich environments. This suggests that the circadian control of mitochondrial dynamics is crucial for preserving metabolic flexibility. Post-transcriptional regulation could provide an extra layer of circadian control over mitochondrial dynamics. The enzyme SIRT3 removes acetyl groups from OPA1, leading to its activation and promotion of fusion (Samant et al. 2014). The activity of SIRT3 relies on the fluctuating amounts of NAD+, as mentioned above.

5.3.3 Signaling Pathways Which Integrate Circadian Biology to Mitochondrial Functions

The molecules Adenosine monophosphate (AMP)-activated protein kinase (AMPK) and mammalian target of rapamycin (mTOR) control metabolic pathways, including autophagy, in opposite ways (Jouffe et al. 2013; Lamia et al. 2009). When there is a scarcity of energy, AMPK stimulates autophagy through ULK1 (Egan et al. 2011). AMPK suppresses the activity of the RAPTOR component of mTOR complex 1 (mTORC1) by phosphorylation. According to Alers et al. (2012), mTORC1 often suppresses autophagy. Toyama et al. present evidence demonstrating that the activation of AMPK triggers the process of mitochondrial fragmentation through the involvement of MFF (Toyama et al. 2016). Furthermore, AMPK phosphorylates and destabilises CRY1, as demonstrated by Lamia et al. in 2009. On the other hand, BMAL1 seems to suppress mTORC1 activity, as indicated by Khapre et al. in 2014. The results indicate that energy-sensing pathways can influence how mitochondria behave, both through direct mechanisms and communication with the molecular clock.

It is well known that circadian rhythmicity and metabolism are correlated (Reinke and Asher 2019); similarly, gut microbiota and the immune system are associated with circadian rhythmicity (Liang et al. 2015; Lu et al. 2017; Scheiermann et al. 2018). Furthermore, there is strong evidence of a connection between the gut microbiota and the immune response (Postler and Ghosh 2017) as well as between metabolism and the immune response (Ganeshan and Chawla 2014). A limited number of studies investigate the interaction between more than two biological domains. When these studies do occur, they usually focus on three domains, such as the interplay between the immune system, metabolism, and circadian rhythmicity (Carroll et al. 2019; Early and Curtis 2016) or between intestinal microbiota, circadian rhythmicity, and metabolism (Parkar et al. 2019).

Using a wide range of scientific literature, we aim to show that the circadian clock, metabolism (at both the systemic and cellular levels), intestinal microbiota, and immune system communicate simultaneously. Furthermore, drawing from recent discoveries in fundamental mitochondrial biology and the role of mitochondria in the four specified biological domains, we suggest that mitochondria control the bidirectional flow of information between each domain. It is important to note that a specific disease cannot be attributed to a single gene or a single biological domain, such as circadian rhythmicity. This is because multiple biological domains are interconnected and rely on each other. Thus, pinpointing a central component in the network, which is the circadian clock, metabolism, intestinal microbiota, and immune system, will enhance our comprehension of their inherent interactions. We suggest that mitochondria serve as this central or "hub" component. Mitochondria are crucial for maintaining energy balance in eukaryotic organisms and play a pivotal role in various physiological processes. Significantly, mitochondria generate signalling molecules and receive and analyse signals from external sources, supporting their suggested role as an "integrative centre". Mitochondrial signals such as mROS, oxidised mtDNA, and extracellular ATP, which are all characteristic signs of malfunctioning mitochondria, trigger the activation of the NLRP3 inflammasome (Próchnicki and Latz 2017). In addition, in conditions characterised by inflammation such as inflammatory bowel disease (IBD), mitochondrial DNA can be released into the bloodstream and identified as a damage-associated molecular pattern (DAMP), so contributing to a broader inflammatory response throughout the body (Boyapati et al. 2018). According to Ip et al. (2017), the absence of IL-10 in individuals with inflammatory bowel disease (IBD) and in mice with colitis leads to the buildup of impaired mitochondria. This, in turn, triggers the activation of the NLRP3 inflammasome in macrophages (Ip et al. 2017). Conversely, the levels of HtrA2 increase in reaction to various stresses (Baker et al. 2011; Moisoi et al. 2009). The study by Peek et al. (2013) demonstrates the connection between NAD+ and the synchronisation of circadian rhythms and metabolism. The level of NAD+ in mitochondria controls the function of sirtuin deacetylase-3 (SIRT3), which manages the acetylation and function of other crucial metabolic enzymes. SIRT3 has the ability to reinstate the function of protein substrates by eliminating acetyl moieties from them (Peterson et al. 2018). The inherent acetyl-transferase function of the circadian regulator CLOCK allows for clock-dependent modification of chromatin structure by adding acetyl groups to histones and non-histone proteins. The deacetylase SIRT1 counterbalances this action, and its effectiveness is contingent upon the amounts of intracellular NAD+ (Nakahata et al. 2009). In the same way, the regular changes in NAD+ levels throughout the day enhance the functions of three specific sirtuins, namely SIRT1, SIRT3, and SIRT6. These sirtuins have a role in regulating several metabolic functions, including glucose, cholesterol, and fatty acid

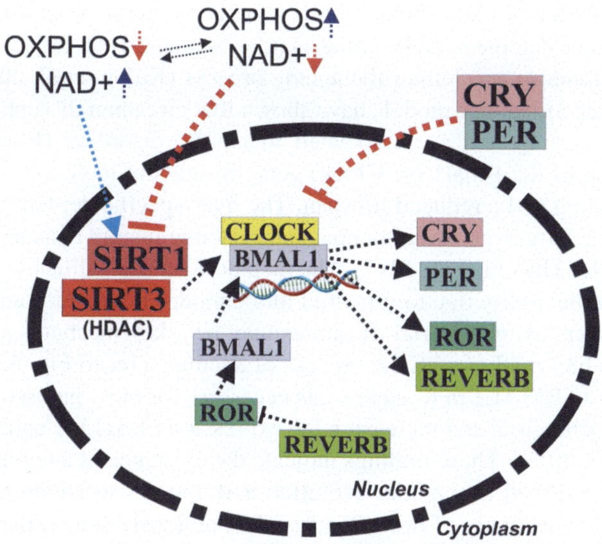

Fig. 5.3 Circadian rhythmicity is regulated by mitochondria via NAD+ production, SIRT1 and SIRT3 activation, and mitochondrial dynamics. The activity of HDACs Sirtuin 1 (SIRT1) and Sirtuin 3 (SIRT3) is NAD+-dependent, and SIRT1 and SIRT3 counteract CLOCK; the synthesis of NAD+ is regulated by circadian rhythmicity, which is associated with mitochondrial dynamics. CLOCK, circadian locomotor output cycles kaput; NAD+, nicotinamide adenin dinucleotide (oxidised); SIRT1, Sirtuin 1; SIRT3, Sirtuin 3; BMAL1, brain and muscle aryl hydrocarbon receptor nuclear translocator-like 1; RORα, retinoic acid receptor-related orphan receptor α; REV-ERBα

metabolism (Imai and Guarente 2016). The levels of various metabolites associated with mitochondria, including NAD+, ATP, mROS, and Krebs cycle intermediates, are influenced by circadian rhythms (Fig. 5.3). Notably, this includes components of the pyruvate dehydrogenase complex, which plays a crucial role in the first step of mitochondrial carbohydrate metabolism, and carnitine palmitoyl-transferase 1, the enzyme responsible for transporting fatty acids into the mitochondrial matrix (Neufeld-Cohen et al. 2016). Aside from the regulation of mitochondrial enzyme levels, mitochondrial respiration is significantly impacted by both the molecular circadian clock and the type of nutrition consumed (Neufeld-Cohen et al. 2016; Scrima et al. 2016). This connection between mitochondrial function, metabolism, and circadian rhythmicity highlights the interplay between these biological processes.

5.4 Circadian Dysregulation and Age-Related Metabolic Diseases: The Role of Mitochondrial Dynamics

Ageing, obesity, and metabolic disorders associated with ageing all seem to have been associated with instances of mitochondrial malfunction. The question of whether this is a cause or effect is still up for debate. Nevertheless, the electron transport chain becomes less effective in ageing cells due to an escalation in electron leakage and a decline in ATP synthesis (Green et al. 2011). Inducing mitochondrial malfunction also accelerates the ageing process (Kujoth et al. 2005; Vermulst et al. 2008). Genetic mouse models have shown that circadian disruption is responsible for both ageing and age-associated metabolic disorders (Kondratov et al. 2006). Deletion of the Bmal1 (or Clock) gene throughout the entire body leads to accelerated ageing and a reduced lifespan. The liver-specific depletion of Bmal1 is enough to cause fatty liver, dyslipidemia, and systemic insulin resistance, as seen in LBmal1KO mice (Jacobi et al. 2017). The metabolic abnormalities seen are caused by metabolic inflexibility due to impaired mitochondrial dynamics and heightened oxidative damage. Mitochondrial dynamics may be a key mechanism that directly connects circadian rhythm with the process of ageing (refer to Fig. 5.4). AHA-1, a protein similar to BMAL1 in *C. elegans*, is necessary for the organised arrangement of muscle mitochondria, and increasing the expression of AHA-1 enhances longevity (Jacobi et al. 2017). These findings indicate the existence of a regulatory mechanism that is conserved throughout evolution and appears to connect longevity or healthspan to the accuracy of mitochondria. As previously stated, the processes of mitochondrial fusion, fission, and mitophagy are essential for managing oxidative stress and maintaining the quality of mitochondria. Uncontrolled production of reactive oxygen species (ROS) leads to the oxidation of proteins and lipids in the mitochondria, causing changes in the arrangement of electron transport chain complexes and the lipid makeup of the mitochondrial membrane. Ageing may be influenced by a combination of decreased mitochondrial biogenesis and clearance, leading to increased mitochondrial damage and inadequate turnover (López-Otín et al. 2013). mtDNA is more prone to mutations compared to nuclear DNA because it lacks histones, has less effective repair mechanisms, and is located close to the electron transport chain, where reactive oxygen species are generated (Linnane et al. 1992). The role of mtDNA mutations in the process of ageing has been a topic

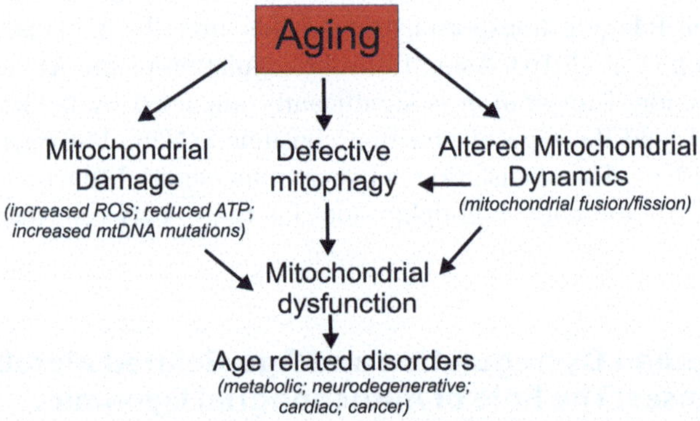

Fig. 5.4 Age-related illnesses and the decline in mitochondrial activity. The process of ageing is linked to a gradual decline in mitochondrial function. Decreases in the functioning of mitochondrial dynamics and quality control are connected to the development of several age-related ailments, such as metabolic syndrome, neurological and cardiovascular diseases, and cancer

of discussion (Park and Larsson 2011). However, it is known that mtDNA mutations in protein or rRNA/tRNA coding areas are linked to multisystem illnesses that exhibit similarities to the ageing process. Furthermore, mice that do not have mitochondrial DNA polymerase G (POLG), which is also referred to as mtDNA mutator mice (Edgar et al. 2009; Hiona et al. 2010), gather mtDNA point mutations/deletions and display premature ageing and a reduced lifespan that is not influenced by ROS (Kujoth et al. 2005; Vermulst et al. 2008).

Multiple research has provided evidence supporting the idea of linking mitochondrial dynamics to the process of ageing. In yeast, the mitochondrial network of double mutants for fusion and fission resembles that of the wild type, but their mitochondrial function is modified and their replicative lifespan is reduced (Bernhardt et al. 2015). On the other hand, genetic methods that encourage the merging of cells (or prevent the splitting of cells) lead to an enhanced ability to withstand stress and longer periods of both chronological and replicative lifespan. Additionally, it was discovered that reducing the activity of DRP1 in *C. elegans* intensified the impact of reduced insulin signalling on lifespan. The lifetime of fruit flies was drastically reduced by mutations in Opa1 or Rho-7, both of which are essential for mitochondrial fusion (McQuibban et al. 2006). Remarkably, Parkin's widespread or neuron-specific amplification in adult fruit flies leads to a longer lifetime (Rana et al. 2013), most likely due to an augmentation in mitophagy. Genetically or pharmacologically enhancing autophagy has been shown to effectively slow down the ageing process in various animal models. The regulatory feedback mentioned is crucial for enhancing stress resistance and promoting longevity (Palikaras et al. 2015). The validity of the mitochondrial theory of ageing was assessed through the generation of knock-in mice, known as "mutator mice", who express a form of the mitochondrial DNA polymerase (POLG) that lacks proofreading capabilities. The mutator mice displayed elevated levels (3–8 times) of mtDNA mutations and premature ageing characteristics such as reduced lifespan, weight loss, hair loss, and spinal curvature, without any concurrent increase in ROS generation. The premature ageing symptoms were not attributed to an elevation in oxidative stress coupled with a buildup of mtDNA mutations, despite the prevailing notion that reactive oxygen species (ROS) is the main driver of ageing (Loeb et al. 2005).

Despite the substitution of the identical amino acid residue, the mutant mice created by the two groups exhibit diverse abnormalities. Initially, Kujoth and colleagues observed increased amounts of caspase-3, an important regulator of programmed cell death, and TUNEL-positive cells (cells labelled with terminal deoxynucleotidyl transferase-mediated deoxyuridine triphosphate nick end labelling) in the tissues of mutator mice. The researchers proposed that an elevated rate of apoptosis was the cause of accelerated ageing characteristics (Kujoth et al. 2005). However, Trifunovic and colleagues showed that mouse embryonic fibroblasts (MEFs) did not exhibit increased sensitivity to apoptosis caused by oxidative stress (Trifunovic et al. 2004). The study also established a correlation between respiratory insufficiency and premature ageing characteristics. Given the high occurrence of mitochondrial interaction, it is expected that random mutations in mutator mice would not lead to respiratory insufficiency.

5.5 Summary

The activity of mitochondria is controlled by the circadian clock through the utilisation of a number of different regulatory mechanisms. Controlling fusion, fission, and mitophagy in a coordinated manner ensures that reactive oxygen species (ROS), which are a consequence of respiration, are efficiently regulated and that damaged components are repaired. It is possible that the duties of mitochondrial biogenesis, energetics, and quality control are in conflict with one another. A method for the adaptation and sequential activation of both processes is provided by the circadian clock's temporal regulation. This system allows for the optimisation of metabolic flexibility, the minimization of oxidative stress, and the preservation of mitochondrial integrity. The dysregulation of BMAL1/CLOCK activity, which occurs in conjunction with the circadian cycle, leads to a decrease in mitochondrial activity and fidelity, which most likely contributes to the process of ageing. In order to fully understand how mitochondrial dynamics contributes to the ageing of cells and organisms, it is necessary to first understand the specific mechanisms involved.

In the future, however, the discovery of mechanisms that either restore or increase the adaptability of mitochondrial dynamics may result in the development of novel therapeutic techniques for the treatment of metabolic diseases that are associated with ageing.

References

Alers S, Löffler AS, Wesselborg S, Stork B (2012) Role of AMPK-mTOR-Ulk1/2 in the regulation of autophagy: cross talk, shortcuts, and feedbacks. Mol Cell Biol 32(1):2–11. https://doi.org/10.1128/MCB.06159-11

Alexander C, Votruba M, Pesch UEA, Thiselton DL, Mayer S, Moore A, Rodriguez M, Kellner U, Leo-Kottler B, Auburger G, Bhattacharya SS, Wissinger B (2000) OPA1, encoding a dynamin-related GTPase, is mutated in autosomal dominant optic atrophy linked to chromosome 3q28. Nat Genet 26(2):211–215. https://doi.org/10.1038/79944

Atger F, Gobet C, Marquis J, Martin E, Wang J, Weger B, Lefebvre G, Descombes P, Naef F, Gachon F (2015) Circadian and feeding rhythms differentially affect rhythmic mRNA transcription and translation in mouse liver. Proc Natl Acad Sci 112(47):E6579. https://doi.org/10.1073/pnas.1515308112

Baker MJ, Tatsuta T, Langer T (2011) Quality control of mitochondrial proteostasis. Cold Spring Harb Perspect Biol 3(7):a007559. https://doi.org/10.1101/cshperspect.a007559

Bass J, Takahashi JS (2010) Circadian integration of metabolism and energetics. Science 330(6009):1349–1354. https://doi.org/10.1126/science.1195027

Bernhardt D, Müller M, Reichert AS, Osiewacz HD (2015) Simultaneous impairment of mitochondrial fission and fusion reduces mitophagy and shortens replicative lifespan. Sci Rep 5(1):7885. https://doi.org/10.1038/srep07885

Boyapati RK, Dorward DA, Tamborska A, Kalla R, Ventham NT, Doherty MK, Whitfield PD, Gray M, Loane J, Rossi AG, Satsangi J, Ho G (2018) Mitochondrial DNA is a pro-inflammatory damage-associated molecular pattern released during active IBD. Inflamm Bowel Dis 24(10):2113–2122. https://doi.org/10.1093/ibd/izy095

Carroll RG, Timmons GA, Cervantes-Silva MP, Kennedy OD, Curtis AM (2019) Immunometabolism around the clock. Trends Mol Med 25(7):612–625. https://doi.org/10.1016/j.molmed.2019.04.013

Chan DC (2006) Mitochondria: dynamic organelles in disease, aging, and development. Cell 125(7):1241–1252. https://doi.org/10.1016/j.cell.2006.06.010

Chang C, Blackstone C (2010) Dynamic regulation of mitochondrial fission through modification of the dynamin-related protein Drp1. Ann N Y Acad Sci 1201(1):34–39. https://doi.org/10.1111/j.1749-6632.2010.05629.x

Chen Y, Dorn GW (2013) PINK1-phosphorylated mitofusin 2 is a Parkin receptor for culling damaged mitochondria. Science 340(6131):471–475. https://doi.org/10.1126/science.1231031

Chen H, Vermulst M, Wang YE, Chomyn A, Prolla TA, McCaffery JM, Chan DC (2010) Mitochondrial fusion is required for mtDNA stability in skeletal muscle and tolerance of mtDNA mutations. Cell 141(2):280–289. https://doi.org/10.1016/j.cell.2010.02.026

Clark IE, Dodson MW, Jiang C, Cao JH, Huh JR, Seol JH, Yoo SJ, Hay BA, Guo M (2006) Drosophila pink1 is required for mitochondrial function and interacts genetically with parkin. Nature 441(7097):1162–1166. https://doi.org/10.1038/nature04779

Early JO, Curtis AM (2016) Immunometabolism: is it under the eye of the clock? Semin Immunol 28(5):478–490. https://doi.org/10.1016/j.smim.2016.10.006

Edgar D, Shabalina I, Camara Y, Wredenberg A, Calvaruso MA, Nijtmans L, Nedergaard J, Cannon B, Larsson N-G, Trifunovic A (2009) Random point mutations with major effects on protein-coding genes are the driving force behind premature aging in mtDNA mutator mice. Cell Metab 10(2):131–138. https://doi.org/10.1016/j.cmet.2009.06.010

Egan D, Kim J, Shaw RJ, Guan K-L (2011) The autophagy initiating kinase ULK1 is regulated via opposing phosphorylation by AMPK and mTOR. Autophagy 7(6):643–644. https://doi.org/10.4161/auto.7.6.15123

Egner A, Jakobs S, Hell SW (2002) Fast 100-nm resolution three-dimensional microscope reveals structural plasticity of mitochondria in live yeast. Proc Natl Acad Sci 99(6):3370–3375. https://doi.org/10.1073/pnas.052545099

Ganeshan K, Chawla A (2014) Metabolic regulation of immune responses. Annu Rev Immunol 32(1):609–634. https://doi.org/10.1146/annurev-immunol-032713-120236

Glick D, Zhang W, Beaton M, Marsboom G, Gruber M, Simon MC, Hart J, Dorn GW, Brady MJ, Macleod KF (2012) BNip3 regulates mitochondrial function and lipid metabolism in the liver. Mol Cell Biol 32(13):2570–2584. https://doi.org/10.1128/MCB.00167-12

Gomes LC, Benedetto GD, Scorrano L (2011) During autophagy mitochondria elongate, are spared from degradation and sustain cell viability. Nat Cell Biol 13(5):589–598. https://doi.org/10.1038/ncb2220

Gong C, Li C, Qi X, Song Z, Wu J, Hughes ME, Li X (2015) The daily rhythms of mitochondrial gene expression and oxidative stress regulation are altered by aging in the mouse liver. Chronobiol Int 32(9):1254–1263. https://doi.org/10.3109/07420528.2015.1085388

Green DR, Galluzzi L, Kroemer G (2011) Mitochondria and the autophagy–inflammation–cell death axis in organismal aging. Science 333(6046):1109–1112. https://doi.org/10.1126/science.1201940

Hebert AS, Dittenhafer-Reed KE, Yu W, Bailey DJ, Selen ES, Boersma MD, Carson JJ, Tonelli M, Balloon AJ, Higbee AJ, Westphall MS, Pagliarini DJ, Prolla TA, Assadi-Porter F, Roy S, Denu JM, Coon JJ (2013) Calorie restriction and SIRT3 trigger global reprogramming of the mitochondrial protein acetylome. Mol Cell 49(1):186–199. https://doi.org/10.1016/j.molcel.2012.10.024

Hiona A, Sanz A, Kujoth GC, Pamplona R, Seo AY, Hofer T, Someya S, Miyakawa T, Nakayama C, Samhan-Arias AK, Servais S, Barger JL, Portero-Otín M, Tanokura M, Prolla TA, Leeuwenburgh C (2010) Mitochondrial DNA mutations induce mitochondrial dysfunction, apoptosis and sarcopenia in skeletal muscle of mitochondrial DNA mutator mice. PLoS One 5(7):e11468. https://doi.org/10.1371/journal.pone.0011468

Imai S, Guarente L (2016) It takes two to tango: NAD+ and sirtuins in aging/longevity control. NPJ Aging Mech Dis 2(1):16017. https://doi.org/10.1038/npjamd.2016.17

Ip WKE, Hoshi N, Shouval DS, Snapper S, Medzhitov R (2017) Anti-inflammatory effect of IL-10 mediated by metabolic reprogramming of macrophages. Science 356(6337):513–519. https://doi.org/10.1126/science.aal3535

Jacobi D, Atger F, Lee C-H (2017) Circadian control of mitochondrial dynamics and its implication in aging. In: Jazwinski SM, Belancio VP, Hill SM (eds) Circadian rhythms and their impact on aging, vol 7. Springer International Publishing, Berlin, pp 147–161. https://doi.org/10.1007/978-3-319-64543-8_7

Jastroch M, Hirschberg V, Klingenspor M (2012) Functional characterization of UCP1 in mammalian HEK293 cells excludes mitochondrial uncoupling artefacts and reveals no contribution to basal proton leak. Biochim Biophys Acta 1817(9):1660–1670. https://doi.org/10.1016/j.bbabio.2012.05.014

Jouffe C, Cretenet G, Symul L, Martin E, Atger F, Naef F, Gachon F (2013) The circadian clock coordinates ribosome biogenesis. PLoS Biol 11(1):e1001455. https://doi.org/10.1371/journal.pbio.1001455

Khapre RV, Patel SA, Kondratova AA, Chaudhary A, Velingkaar N, Antoch MP, Kondratov RV (2014) Metabolic clock generates nutrient anticipation rhythms in mTOR signaling. Aging 6(8):675–689. https://doi.org/10.18632/aging.100686

Koike N, Yoo S-H, Huang H-C, Kumar V, Lee C, Kim T-K, Takahashi JS (2012) Transcriptional architecture and chromatin landscape of the core circadian clock in mammals. Science 338(6105):349–354. https://doi.org/10.1126/science.1226339

Kojima S, Sher-Chen EL, Green CB (2012) Circadian control of mRNA polyadenylation dynamics regulates rhythmic protein expression. Genes Dev 26(24):2724–2736. https://doi.org/10.1101/gad.208306.112

Kondratov RV, Kondratova AA, Gorbacheva VY, Vykhovanets OV, Antoch MP (2006) Early aging and age-related pathologies in mice deficient in BMAL1, the core component of the circadian clock. Genes Dev 20(14):1868–1873. https://doi.org/10.1101/gad.1432206

Koyano F, Okatsu K, Kosako H, Tamura Y, Go E, Kimura M, Kimura Y, Tsuchiya H, Yoshihara H, Hirokawa T, Endo T, Fon EA, Trempe J-F, Saeki Y, Tanaka K, Matsuda N (2014) Ubiquitin is phosphorylated by PINK1 to activate parkin. Nature 510(7503):162–166. https://doi.org/10.1038/nature13392

Kujoth GC, Hiona A, Pugh TD, Someya S, Panzer K, Wohlgemuth SE, Hofer T, Seo AY, Sullivan R, Jobling WA, Morrow JD, Van Remmen H, Sedivy JM, Yamasoba T, Tanokura M, Weindruch R, Leeuwenburgh C, Prolla TA (2005) Mitochondrial DNA mutations, oxidative stress, and apoptosis in mammalian aging. Science 309(5733):481–484. https://doi.org/10.1126/science.1112125

Lamia KA, Sachdeva UM, DiTacchio L, Williams EC, Alvarez JG, Egan DF, Vasquez DS, Juguilon H, Panda S, Shaw RJ, Thompson CB, Evans RM (2009) AMPK regulates the circadian clock by cryptochrome phosphorylation and degradation. Science 326(5951):437–440. https://doi.org/10.1126/science.1172156

Liang X, Bushman FD, FitzGerald GA (2015) Rhythmicity of the intestinal microbiota is regulated by gender and the host circadian clock. Proc Natl Acad Sci 112(33):10479–10484. https://doi.org/10.1073/pnas.1501305112

Liesa M, Shirihai OS (2013) Mitochondrial dynamics in the regulation of nutrient utilization and energy expenditure. Cell Metab 17(4):491–506. https://doi.org/10.1016/j.cmet.2013.03.002

Linnane AW, Zhang C, Baumer A, Nagley P (1992) Mitochondrial DNA mutation and the ageing process: bioenergy and pharmacological intervention. Mutat Res 275(3–6):195–208. https://doi.org/10.1016/0921-8734(92)90023-I

Loeb LA, Wallace DC, Martin GM (2005) The mitochondrial theory of aging and its relationship to reactive oxygen species damage and somatic mtDNA mutations. Proc Natl Acad Sci 102(52):18769–18770. https://doi.org/10.1073/pnas.0509776102

López-Otín C, Blasco MA, Partridge L, Serrano M, Kroemer G (2013) The hallmarks of aging. Cell 153(6):1194–1217. https://doi.org/10.1016/j.cell.2013.05.039

Losón OC, Song Z, Chen H, Chan DC (2013) Fis1, Mff, MiD49, and MiD51 mediate Drp1 recruitment in mitochondrial fission. Mol Biol Cell 24(5):659–667. https://doi.org/10.1091/mbc.e12-10-0721

Lu H, McClung CR, Zhang C (2017) Tick tock: circadian regulation of plant innate immunity. Annu Rev Phytopathol 55(1):287–311. https://doi.org/10.1146/annurev-phyto-080516-035451

Marcheva B, Ramsey KM, Peek CB, Affinati A, Maury E, Bass J (2013) Circadian clocks and metabolism. In: Kramer A, Merrow M (eds) Circadian clocks, vol 217. Springer, Berlin, pp 127–155. https://doi.org/10.1007/978-3-642-25950-0_6

McQuibban GA, Lee JR, Zheng L, Juusola M, Freeman M (2006) Normal mitochondrial dynamics requires rhomboid-7 and affects drosophila lifespan and neuronal function. Curr Biol 16(10):982–989. https://doi.org/10.1016/j.cub.2006.03.062

Mitchell P (1961) Coupling of phosphorylation to electron and hydrogen transfer by a chemiosmotic type of mechanism. Nature (London) 191:105–105

Moisoi N, Klupsch K, Fedele V, East P, Sharma S, Renton A, Plun-Favreau H, Edwards RE, Teismann P, Esposti MD, Morrison AD, Wood NW, Downward J, Martins LM (2009) Mitochondrial dysfunction triggered by loss of HtrA2 results in the activation of a brain-specific transcriptional stress response. Cell Death Differ 16(3):449–464. https://doi.org/10.1038/cdd.2008.166

Moullan N, Mouchiroud L, Wang X, Ryu D, Williams EG, Mottis A, Jovaisaite V, Frochaux MV, Quiros PM, Deplancke B, Houtkooper RH, Auwerx J (2015) Tetracyclines disturb mitochondrial function across eukaryotic models: a call for caution in biomedical research. Cell Rep 10(10):1681–1691. https://doi.org/10.1016/j.celrep.2015.02.034

Nakada K, Inoue K, Ono T, Isobe K, Ogura A, Goto Y-I, Nonaka I, Hayashi J-I (2001) Inter-mitochondrial complementation: mitochondria-specific system preventing mice from expression of disease phenotypes by mutant mtDNA. Nat Med 7(8):934–940. https://doi.org/10.1038/90976

Nakahata Y, Sahar S, Astarita G, Kaluzova M, Sassone-Corsi P (2009) Circadian control of the NAD^+ salvage pathway by CLOCK-SIRT1. Science 324(5927):654–657. https://doi.org/10.1126/science.1170803

Neufeld-Cohen A, Robles MS, Aviram R, Manella G, Adamovich Y, Ladeuix B, Nir D, Rousso-Noori L, Kuperman Y, Golik M, Mann M, Asher G (2016) Circadian control of oscillations in mitochondrial rate-limiting enzymes and nutrient utilization by PERIOD proteins. Proc Natl Acad Sci 113(12):E1673. https://doi.org/10.1073/pnas.1519650113

Nobes CD, Brown GC, Olive PN, Brand MD (1990) Non-ohmic proton conductance of the mitochondrial inner membrane in hepatocytes. J Biol Chem 265(22):12903–12909. https://doi.org/10.1016/S0021-9258(19)38245-6

Palikaras K, Lionaki E, Tavernarakis N (2015) Coordination of mitophagy and mitochondrial biogenesis during ageing in C. elegans. Nature 521(7553):525–528. https://doi.org/10.1038/nature14300

Park CB, Larsson N-G (2011) Mitochondrial DNA mutations in disease and aging. J Cell Biol 193(5):809–818. https://doi.org/10.1083/jcb.201010024

Parkar S, Kalsbeek A, Cheeseman J (2019) Potential role for the gut microbiota in modulating host circadian rhythms and metabolic health. Microorganisms 7(2):41. https://doi.org/10.3390/microorganisms7020041

Peek CB, Affinati AH, Ramsey KM, Kuo H-Y, Yu W, Sena LA, Ilkayeva O, Marcheva B, Kobayashi Y, Omura C, Levine DC, Bacsik DJ, Gius D, Newgard CB, Goetzman E, Chandel NS, Denu JM, Mrksich M, Bass J (2013) Circadian clock NAD^+ cycle drives mitochondrial oxidative metabolism in mice. Science 342(6158):1243417. https://doi.org/10.1126/science.1243417

Peterson BS, Campbell JE, Ilkayeva O, Grimsrud PA, Hirschey MD, Newgard CB (2018) Remodeling of the acetylproteome by SIRT3 manipulation fails to affect insulin secretion or β cell metabolism in the absence of overnutrition. Cell Rep 24(1):209–223.e6. https://doi.org/10.1016/j.celrep.2018.05.088

Pimenta De Castro I, Costa AC, Lam D, Tufi R, Fedele V, Moisoi N, Dinsdale D, Deas E, Loh SHY, Martins LM (2012) Genetic analysis of mitochondrial protein misfolding in Drosophila melanogaster. Cell Death Differ 19(8):1308–1316. https://doi.org/10.1038/cdd.2012.5

Postler TS, Ghosh S (2017) Understanding the holobiont: how microbial metabolites affect human health and shape the immune system. Cell Metab 26(1):110–130. https://doi.org/10.1016/j.cmet.2017.05.008

Próchnicki T, Latz E (2017) Inflammasomes on the crossroads of innate immune recognition and metabolic control. Cell Metab 26(1):71–93. https://doi.org/10.1016/j.cmet.2017.06.018

Ramsey KM, Yoshino J, Brace CS, Abrassart D, Kobayashi Y, Marcheva B, Hong H-K, Chong JL, Buhr ED, Lee C, Takahashi JS, Imai S, Bass J (2009) Circadian clock feedback cycle through NAMPT-mediated NAD^+ biosynthesis. Science 324(5927):651–654. https://doi.org/10.1126/science.1171641

Rana A, Rera M, Walker DW (2013) Parkin overexpression during aging reduces proteotoxicity, alters mitochondrial dynamics, and extends lifespan. Proc Natl Acad Sci 110(21):8638–8643. https://doi.org/10.1073/pnas.1216197110

Reinke H, Asher G (2019) Crosstalk between metabolism and circadian clocks. Nat Rev Mol Cell Biol 20(4):227–241. https://doi.org/10.1038/s41580-018-0096-9

Rojo M, Legros F, Chateau D, Lombès A (2002) Membrane topology and mitochondrial targeting of mitofusins, ubiquitous mammalian homologs of the transmembrane GTPase Fzo. J Cell Sci 115(8):1663–1674. https://doi.org/10.1242/jcs.115.8.1663

Samant SA, Zhang HJ, Hong Z, Pillai VB, Sundaresan NR, Wolfgeher D, Archer SL, Chan DC, Gupta MP (2014) SIRT3 deacetylates and activates OPA1 to regulate mitochondrial dynamics during stress. Mol Cell Biol 34(5):807–819. https://doi.org/10.1128/MCB.01483-13

Santel A, Fuller MT (2001) Control of mitochondrial morphology by a human mitofusin. J Cell Sci 114(5):867–874. https://doi.org/10.1242/jcs.114.5.867

Scheiermann C, Gibbs J, Ince L, Loudon A (2018) Clocking in to immunity. Nat Rev Immunol 18(7):423–437. https://doi.org/10.1038/s41577-018-0008-4

Schon EA, Gilkerson RW (2010) Functional complementation of mitochondrial DNAs: mobilizing mitochondrial genetics against dysfunction. Biochim Biophys Acta Gen Subj 1800(3):245–249. https://doi.org/10.1016/j.bbagen.2009.07.007

Scrima R, Cela O, Merla G, Augello B, Rubino R, Quarato G, Fugetto S, Menga M, Fuhr L, Relógio A, Piccoli C, Mazzoccoli G, Capitanio N (2016) Clock-genes and mitochondrial respiratory activity: evidence of a reciprocal interplay. Biochim Biophys Acta 1857(8):1344–1351. https://doi.org/10.1016/j.bbabio.2016.03.035

Tanaka A, Cleland MM, Xu S, Narendra DP, Suen D-F, Karbowski M, Youle RJ (2010) Proteasome and p97 mediate mitophagy and degradation of mitofusins induced by Parkin. J Cell Biol 191(7):1367–1380. https://doi.org/10.1083/jcb.201007013

Toyama EQ, Herzig S, Courchet J, Lewis TL, Losón OC, Hellberg K, Young NP, Chen H, Polleux F, Chan DC, Shaw RJ (2016) AMP-activated protein kinase mediates mitochondrial fission in response to energy stress. Science 351(6270):275–281. https://doi.org/10.1126/science.aab4138

Trifunovic A, Wredenberg A, Falkenberg M, Spelbrink JN, Rovio AT, Bruder CE, Bohlooly-Y M, Gidlöf S, Oldfors A, Wibom R, Törnell J, Jacobs HT, Larsson N-G (2004) Premature ageing in mice expressing defective mitochondrial DNA polymerase. Nature 429(6990):417–423. https://doi.org/10.1038/nature02517

Valente EM, Abou-Sleiman PM, Caputo V, Muqit MMK, Harvey K, Gispert S, Ali Z, Del Turco D, Bentivoglio AR, Healy DG, Albanese A, Nussbaum R, González-Maldonado R, Deller T, Salvi S, Cortelli P, Gilks WP, Latchman DS, Harvey RJ et al (2004) Hereditary early-onset Parkinson's disease caused by mutations in *PINK1*. Science 304(5674):1158–1160. https://doi.org/10.1126/science.1096284

Vermulst M, Wanagat J, Kujoth GC, Bielas JH, Rabinovitch PS, Prolla TA, Loeb LA (2008) DNA deletions and clonal mutations drive premature aging in mitochondrial mutator mice. Nat Genet 40(4):392–394. https://doi.org/10.1038/ng.95

Westermann B (2010) Mitochondrial fusion and fission in cell life and death. Nat Rev Mol Cell Biol 11(12):872–884. https://doi.org/10.1038/nrm3013

Wikstrom JD, Mahdaviani K, Liesa M, Sereda SB, Si Y, Las G, Twig G, Petrovic N, Zingaretti C, Graham A, Cinti S, Corkey BE, Cannon B, Nedergaard J, Shirihai OS (2014) Hormone-induced mitochondrial fission is utilized by brown adipocytes as an amplification pathway for energy expenditure. EMBO J 33:418. https://doi.org/10.1002/embj.201385014

Youle RJ, Narendra DP (2011) Mechanisms of mitophagy. Nat Rev Mol Cell Biol 12(1):9–14. https://doi.org/10.1038/nrm3028

Youle RJ, Van Der Bliek AM (2012) Mitochondrial fission, fusion, and stress. Science 337(6098):1062–1065. https://doi.org/10.1126/science.1219855

Zhao Y, Ye L, Liu H, Xia Q, Zhang Y, Yang X, Wang K (2010) Vanadium compounds induced mitochondria permeability transition pore (PTP) opening related to oxidative stress. J Inorg Biochem 104(4):371–378. https://doi.org/10.1016/j.jinorgbio.2009.11.007

Health Consequences of Circadian Disruption in Humans and Animal Models During Aging

6

6.1 Introduction

The circadian system, which controls a variety of rhythms in our bodies, is under a lot of stress in modern times. With the demand for products and services expanding throughout the day, rapid industrialisation and globalisation of the economy, and nontraditional work schedules are becoming increasingly prevalent. Employees who work shifts that involve the night must stay awake when their biological clocks signal inactivity. A growing body of research suggests that shift work raises the risk of several major physiological and psychological conditions, including cardiovascular disease, metabolic syndrome, obesity, immunological dysfunction, cancer, and problems during pregnancy. Due to survivorship bias and self-selection, epidemiological studies may understate shift employment's negative effects on health.

The three main reasons that are believed to have a negative impact on a shift worker's performance and health are (1) circadian misalignment, (2) lack of sleep, and (3) exposure to light at night. Nevertheless, the presence of different shift exposure settings and the potential influence of influencing variables (such as changes in diet, social stress, and the consumption of nicotine, alcohol, and prescription drugs) make it difficult to accurately determine the exact importance of circadian disruption in epidemiological studies. There are impacts of circadian disruption seen in epidemiological studies that account for some of these noncircadian characteristics.

These findings, along with experimental data, strongly suggest that circadian disturbance and misalignment are major causes of the negative health effects frequently seen in shift workers. Studies using circadian disruption animal models are beneficial because they allow for stricter control to examine the role of circadian misalignment alone or in combination with these other parameters.

The extensive impacts of circadian disturbance on human and animal models' health are outlined in this chapter. The circadian system and various forms of circadian disturbance in humans are first described, along with the appropriate

S. I. Rizvi, G. Majumdar, *Circadian Clock and Aging*,
https://doi.org/10.1007/978-981-97-6695-6_6

experimental procedures that were utilized to mimic these situations in real people and animal models. Second, we give a thorough analysis of the noticeable effects of these circumstances on health and talk about the possible causes of various illness states. Finally, we discuss areas for future research that can pinpoint the precise causes and effects of circadian disturbance and potential remedies.

6.2 Circadian Regulation of Physiology

The suprachiasmatic nucleus (SCN) of the anterior hypothalamus has a master pacemaker that governs the hierarchy of biological clocks that make up the circadian system in mammals. Light synchronises the SCN, which then coordinates the tissues that follow so that they adopt particular temporal relationships to the outside world and to one another. The SCN regulates overt rhythms, including body temperature, eating, and hormone production, in addition to monosynaptic and polysynaptic neuronal projections to downstream destinations, which can work to synchronise peripheral clocks. Molecular oscillators present in individual cells of central and peripheral clocks generate circadian rhythms in protein expression and metabolic state. For the proper functioning of the circadian rhythm, it is essential to synchronise with the 24-h environment and local time and coordinate molecular and systemic processes (Buhr and Takahashi 2013).

A population of intrinsically photoresponsive retinal ganglion cells that express the photopigment melanopsin provide a direct retinal projection to the SCN (Hattar et al. 2006). Acute impacts of nighttime light include alterations in melatonin levels, sleep, alertness, SCN electrical activity, and gene expression. Contrary to the conventional belief that the human circadian visual system has a relatively high photic threshold in comparison to the traditional image-forming visual system, current research suggests that even very low levels of illumination at night can have an impact on the circadian visual system (Chew et al. 2017). This implies that even low amounts of artificial illumination used at home or the office may change circadian function.

Autoregulatory feedback loops that regulate the transcription and translation of clock genes are responsible for regulating circadian rhythms at the molecular level (Fagiani et al. 2022). The period (per1, per2) and cryptochrome (cry1, cry2) genes are activated to produce proteins that interact with casein kinase1e and move back into the nucleus to suppress CLOCK/BMAL1-mediated transcription within the core molecular loop. Every 24 h, the cycle might restart due to the degradation of PER/CRY repressor elements, which releases the inhibition on CLOCK/BMAL1 transcription. This rhythmic mechanism affects nearly all of the body's cells and regulates both general and tissue-specific activity.

The expression of clock genes and proteins is also modulated by additional loops involving other clock gene elements, which affects the amplitude and rate of circadian rhythms (such REV-ERB and ROR). Additionally, to affect cellular activity, these transcriptional-translational feedback loops collaborate with electrical and metabolic oscillations (Jagannath et al. 2017), and each bodily cell has the ability to regulate both general and tissue-specific functions. The clock is modulated by other loops utilising different clock gene components.

6.3 Types of Circadian Disruption in Humans and Animal Models

Despite its frequent use, “circadian disruption” is essentially an umbrella word covering a wide variety of distinct circadian physiological perturbations (Vetter 2020). This includes “desynchrony” when two or more rhythms exhibit a different period and “misalignment” when two or more rhythms adopt an aberrant phase relationship. Additionally, the terms “misalignment” and “desynchrony” can relate to internal (such as central vs. peripheral clocks) or external (such as central vs. external LD cycle) rhythms. Additionally, the term “chronodisruption” is also used to refer to the discrepancy between a person’s biological night and work hours (Erren et al. 2003). The phrase “social jetlag” is commonly employed to describe the daily strain that work and social responsibilities impose on the circadian rhythm. A valuable marker of circadian misalignment, social jetlag is defined as the difference in sleep schedule between work and free days (Wittmann et al. 2006).

6.4 Measuring Circadian Disruption

Different forms of circadian disturbance will result in distinctive adjustments to behaviour patterns. For instance, circadian misalignment will alter the timing of activity’s commencement and phasing relative to the LD cycle [West et al., Nat Comm]. On the other hand, internal desynchrony can lead to two or more rhythms with distinct periodicities, as seen under non-24 h T-cycles (Cambras et al. 2007). In nocturnal rodents, chronodisruption is predicted to result in higher activity during the typical inactive phase (increased activity during the light phase), with a decrease in the relative amplitude between activity and rest. Last but not least, social jetlag will cause activity onset on “work” and “free” days to change. However, because of its connection to human social activity, such paradigms are rarely applied in animal investigations. Some consistent alterations are seen across a wide range of situations, in addition to these recognizable changes in activity seen under various circadian disruption scenarios. When compared to normal, healthy rhythms, the circadian period normally remains intact, but activity rhythms become more fragmented and irregular. As a result, quantifying circadian disruption does not really fit the conventional analytical instruments used in circadian biology, which are tuned to detect changes in the period.

Numerous methods have been employed in the literature to evaluate circadian disturbance. These techniques range from straightforward actogram visual inspection to measurements like periodogram power, variability in activity initiation, light phase activity, activity bouts, intra-daily variability, and relative amplitude. In more detail, these are discussed below with reference to various types of circadian disturbance. We then go over how these measurements of circadian disruption are created and give further information about their advantages and disadvantages. Figure 6.1 displays schematic illustrations of normal and interrupted rhythms for each parameter. The following discussion is based on research on circadian disruption in laboratory animals (rats and mice). However, many of these techniques can be applied to studies of human illnesses and, in some cases, were initially created for such research (see Vetter 2020 for a thorough comparison of experimental human, animal, and observational/field investigations).

6.5 Circadian Disruption in the Real World

One of the main causes of circadian disturbance in the modern world is shift work or nontraditional work. Nontraditional work schedules are those that take place outside of the 9 am to 5 pm timeframe; however, this category includes a wide range of schedules, such as set night shifts, early morning shifts, flex shifts, extended shifts,

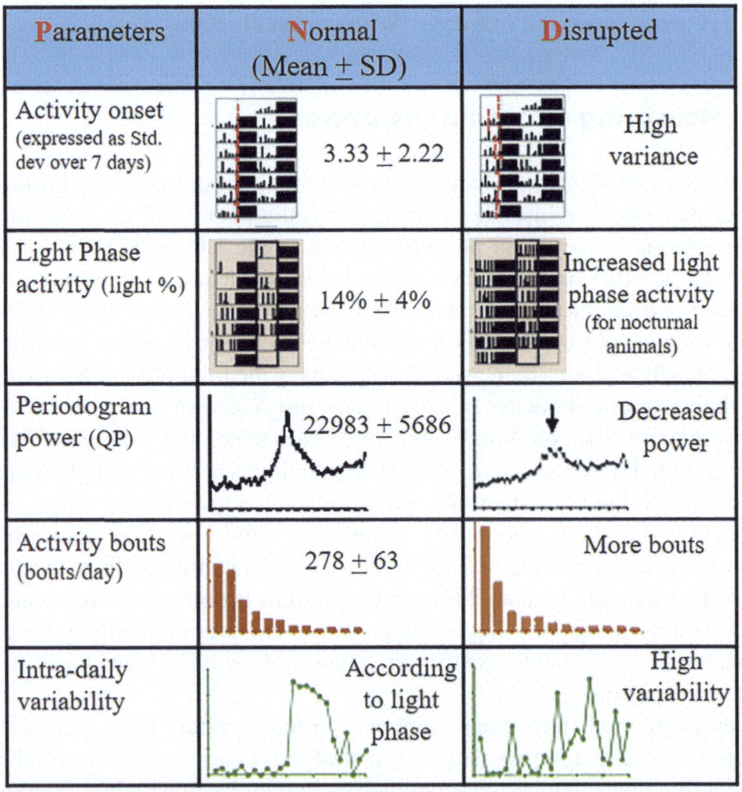

Fig. 6.1 Schematic illustrations of normal and interrupted rhythms for circadian clock related activity parameters

rotating shifts, and regular overseas travel (Costa 1997). According to estimates, more than 20% of US workers (more than 20 million people) work nontraditional hours, which is expected to cost the economy more than $200 billion annually in sick days, stress, lost productivity, and accidents both on and off the job (Kerin and Aguirre 2004; Folkard and Tucker 2003). While shift workers can push themselves to sleep during the day and be up at night, this frequently results in decreased awareness and sleep efficiency (Akerstedt and Wright 2009). A majority of shift workers do not exhibit a corresponding shift in circadian function and instead continue to be in a state where the biological clock is programming sleep during the night and wakefulness during the day, despite the ability to consciously meet expected behavioural demands (Dumont et al. 2012). The next major reason is trans-world travel, which causes jetlag and disrupted clock schedules. Our biological clocks, at some point, adjust to the new time zone after transmeridian travel, but the brief mismatch between internal timing and external conditions results in jetlag symptoms that could continue for days to weeks. The vast and varied symptoms of jetlag are typical of the influence and significance of the circadian rhythm for human health, even though it is sometimes dismissed as a brief inconvenience. The latency to recover is higher when travelling eastward after transmeridian travel, and it normally takes 1 day to acclimatise for each time zone crossed. However, a variety of individual and experience factors can influence the degree and severity of jetlag.

Large-scale population surveys also show that many people experience "social jetlag", which is defined as a mismatch between social demands and the naturally preferred time as determined by an individual's circadian system (i.e., morning larks and evening owls, respectively) (Roenneberg et al. 2012). Finally, disrupted sleep-wake schedules are the major reason for clock disruption. This may be the result of shift working or travelling but also due to sleep disorders. When humans fail to synchronise with the external light:dark (LD) cycle and instead "free-run" with an inherent period other than 24 h, it is known as non-24 h sleep:wake disorder. These individuals' daily rhythms gradually advance or recede each day, which causes sporadic mismatch between internal and external chronology. Non-24 h sleep:wake disorder is most common in people without photic input to the SCN24, while it has also been observed in sighted people. Patients with non-24 h sleep:wake disorder may be able to maintain a consistent daily schedule of awake but also present with mental symptoms, cognitive impairments, and cyclical patterns of daytime sleepiness and overnight alertness (Lockley et al. 2008).

6.6 Models of Circadian Disruption in Humans and Animals

As the trend towards a "24 h" society continues, a growing number of people are becoming more active at night rather than during the day, when our ancestors evolved to be active during the diurnal hours of the 24 h day. A significant portion of the population is affected by some form of circadian disruption (e.g., shift work, jet lag). Although the link between circadian disruption and metabolism has long been observed in humans (e.g., shift workers), the underlying causes of metabolic disease, obesity, diabetes, and cardiometabolic disorders are still poorly understood, necessitating the use of animal models to clarify how circadian dysfunction can result in different pathologies. Most animal studies of circadian disturbance have shifted the LD cycle to mimic human jet lag or shift work. These light-shifting methods are useful for evaluating the detrimental effects of changing the circadian clock. Still, they do not accurately reflect what occurs in people who frequently undergo changes in food patterns to meet work or social situations. Animals on restricted feeding protocols can be used to study the impact of aberrant LD cycles and feeding rhythm alignment. The idea that the timing of feedings alone may have a significant impact on metabolism has been reinforced by recent research. It is crucial to remember that, just as metabolism can influence rhythms, so can rhythms influence metabolism, as is seen when environmental and genetic changes to metabolism result in irregular rhythms. Understanding the biochemical and genetic relationships between the circadian clock and metabolic systems in diverse animal models has opened up new research directions into the processes underlying circadian disturbance and metabolic disease in humans (for details, refer to Arble et al. 2010). Below we detail some specific model conditions used to understand clock disruption and its health effects.

Rotating shift work has been replicated in nocturnal rodent models using the LD cycle (see later), but static night shift work has been replicated using moving wheels, limiting food access, or operant conditioning techniques to impose diurnal work schedules (Salgado-Delgado et al. 2010). The temporal relationship between the SCN and peripheral tissues may change as a result of these protocols, in addition to producing an obvious misalignment with the surrounding environment.

To simulate jet lag in people and rodents, the LD cycle is often moved 4–11 h in one of two directions: eastward (advance) or westward (delay). Humans and rodent models often need more time to acclimatise to LD cycle advancements (Zee et al. 1992). Complete inversion of the LD cycle after a 12-h shift typically results in resynchronization in the delay direction, though species and individual differences may affect this. According to research, simulated jetlag causes a variety of circadian disruptions, such as a brief misalignment of the internal and external milieus, misalignment between the various tissues that make up the circadian system (Sellix et al. 2012), misalignment between neural oscillators within the SCN (Rohling et al. 2011) and misalignment between the molecular parts of the circadian clock (Reddy et al. 2002) Additionally, when given a specific jetlag protocol, a portion of Siberian hamsters either fail to resynchronize or develop arrhythmias rather than doing so (Ruby et al. 1998) Humans (Jewett et al. 1991) and types of nocturnal rodents have both been known to exhibit strange resetting behaviour (Honma and Honma 1999). These unusual resetting behaviours could be an indication that the circadian clock has been slowed down by light during the middle of the subjective night, which is thought to be a phase singularity point in the circadian cycle (Jewett et al. 1999).

In the lab, exposure to non-24 h LD cycles can be used to create a mismatch between internal timing and external cycles (Gronfier et al. 2007) which may be analogous to work patterns used in various occupations (Kelly et al. 1999). Exposure to non-24 h LD cycles outside of these bounds leads rhythms to "free-run", with intermittent and transient alignment between the internal and external milieu. The circadian system can entrain non-24 h cycles only within a specific range (usually 24 ± 2 h). Internal desynchrony, which occurs when several rhythms (such as body temperature, melatonin, and sleep/wake rhythms) free-run independently of one another, can also be caused in humans by exposure to non-24 h LD cycles (Dijk et al. 1999). Last but not least, subjecting rats to non-24-hour LD cycles might lead to the desynchronization of various SCN compartments, leaving one synchronised to the environment's lighting and the other free-running (de la Iglesia et al. 2004).

Exposure to nighttime light is a characteristic of both biological and industrialised environments. Annual variations in day length affect how long people are exposed to daylight, while using artificial light helps mitigate seasonal variations in day length. Even moderate levels of light at night can cause individuals to readjust their circadian rhythms and decrease the release of melatonin. Exposure to continuous light lengthens the circadian period in nocturnal rodents and may result in the "splitting" of behavioural and physiological rhythms (de la Iglesia et al. 2000). Recent research has also examined the impacts of nighttime light pollution by exposing nocturnal rodents to low levels of light (usually 5 lux) during otherwise regular LD cycles (Fonken et al. 2010; Bedrosian et al. 2011). Last but not least, non-24 h LD cycles that do not support entrainment result in aberrant nighttime light exposure, with higher exposure occurring under extremely short LD cycles (such LD cycles of 3.5 h of light and 3.5 h of darkness; LeGates et al. 2012).

Deficits resulting from molecular oscillator mutations can range from a shift in the period to a total absence of circadian synchronisation. The tau mutant hamster, which exhibits a 2 h reduction in period with each copy of the defective gene (Ralph and Menekar 1988) due to a reduction in the phosphorylation capacity of casein kinase 1e, was the first genetic model to be discovered (Lowrey et al. 2000). A number of mice models have been created that exhibit altered circadian clock function thanks to the development of molecular techniques (Siepka et al. 2007). In their most extreme forms, many mouse models—such as the bmal1 knockout mouse, the clock mutant mouse, the per1/per2/mouse, and the cry1/cry2/mouse—are arrhythmic in environments that are always dark. The masking effects of light and darkness allow these mouse models, which lack self-sustaining circadian rhythms, to still exhibit daily variations in behaviour and physiology when kept under 24-h light-dark cycles. A shift in the free-running time and species-atypical alignment to the 24 h LD cycle can result from deletions of just one clock gene paralog, such as per1, albeit this is often inadequate to totally eliminate circadian rhythms. Because the overall clock function is preserved but is altered in a way that disrupts the temporal link with the environment, these latter models may be of particular interest for addressing the effects of circadian disturbance.

6.7 Health Consequences of Circadian Disruption

The SCN has historically been the target of brain injuries in rat models to study the effects of total loss of circadian rhythmicity (Stephan and Zucker 1972). Recent research, however, has concentrated on the genetic and environmental modulation of circadian rhythm. By preserving synchronicity between internal circadian rhythms, behaviours, and environmental stimuli, the circadian system is essential for healthy functioning. But many medical professionals are not completely aware of how important the circadian clock is to human health. Most clinical research on circadian rhythms has been conducted on sleep disorders. However, the circadian system's effects on health are much more extensive. Clinical diagnoses frequently ignore circadian impacts on physiology in favour of daytime assessments at a particular time point. Using (external) clock time misses the significant interindividual variations in internal timing even when time is taken into account. It could soon be possible to include circadian timing in clinical diagnosis and disease therapy thanks to recent improvements in methods for estimating the circadian phase. The circadian system is now understood to be a multioscillator system, despite previously being thought of as a single oscillator. In addition to the hypothalamic suprachiasmatic nucleus' (SCN) primary pacemaker, the body's peripheral oscillators are found in almost all of its organs and cells (Qian and Scheer 2016). Circadian misalignment is the term used to describe irregular timing between peripheral oscillators, SCN rhythms, behaviour (such as sleeping during the biological day), the environment (such as exposure to light during the biological night), or both. Below we describe specific examples of morbidities/diseases/disorders directly affected by circadian disruption (also see Fig. 6.2). The examples are far from being a complete list.

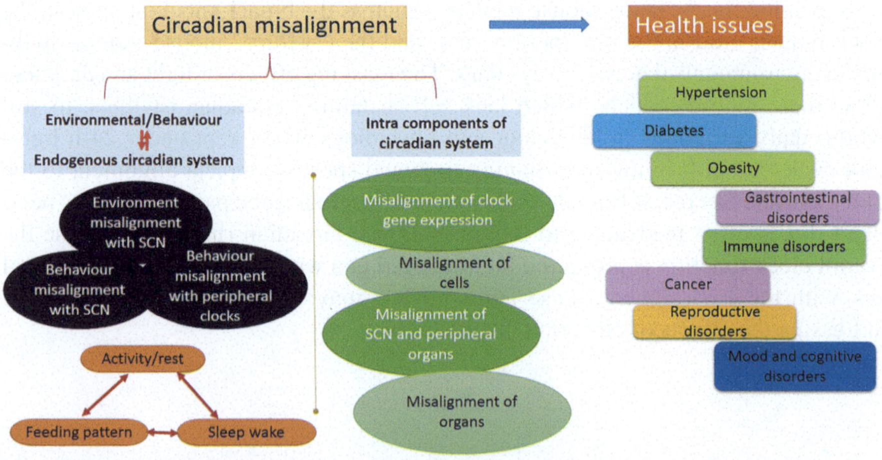

Fig. 6.2 Types of circadian misalignments and specific examples of morbidities/diseases/disorders which are directly caused by the misalignments

6.7.1 Longevity and Aging

Shift workers run a higher risk of dying (Karlsson et al. 2005). Numerous organisms, including cyanobacteria, plants, and flies (Woelfle et al. 2004), have been hypothesized to benefit from optimal alignment to environmental cycles in terms of fitness and lifespan (Aschoff et al. 1971; Pittendrigh and Minis 1972). In this research, housing organisms under LD cycles that are different from the inherent period of the organism results in decreased longevity or fitness. Although the number of comparable research is constrained by animals' relatively long gestation and lifetime, some evidence suggests that circadian misalignment imposes a biological cost that can shorten lifespan and hasten the ageing processes (see Fig. 6.3; Wyse et al. 2010).

Halberg and colleagues were the first to show that mice given weekly LD cycle inversions experienced increased mortality (Halberg and Cadotte 1975). Similar findings were made in a group of Syrian hamsters prone to cardiomyopathy: weekly LD cycle inversions resulted in an 11% reduction in median lifetime (Penev et al. 1998). Since continual light can have therapeutic effects in cardiomyopathic hamsters, it was hypothesized that this result was due to circadian disruption rather than increased light exposure (Natelson et al. 1996). The LD cycle's weekly 6 h changes can also have a significant impact on lifespan, with older mice dying off at a greater rate despite no rise in faecal corticosterone (Davidson et al. 2006). Simulated seasonal variations can also have an impact on lifespan; mouse lemurs kept in accelerated seasonal cycles showed a 30% decrease in average lifetime (Perret 1997).

In a similar way, flies lived shorter lives when the length of the day changed quickly and frequently (le et al. 1997).

Despite earlier claims that health was not adversely affected, recent investigations have shown that molecular impairments in circadian clock operation diminish survival and speed up ageing (Bunger et al. 2000). When heterozygote tau mutant hamsters are maintained under static 24 h LD cycles to which they entrain in a species-atypical way, a 38% reduction in median lifespan with broad organ failure occurs (Hurd and Ralph 1998). It is interesting to note that non-entrainment in

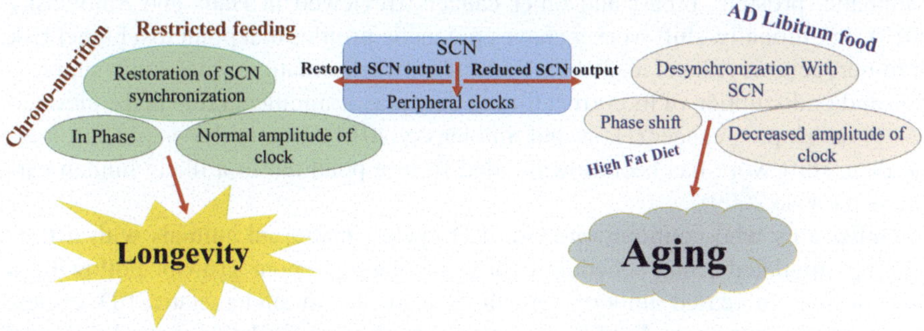

Fig. 6.3 Schematic representation of how circadian misalignment may influence longevity or hasten ageing

homozygote tau mutant hamsters did not result in significantly higher mortality (9% decrease in median lifespan), leading to the hypothesis that misalignment (such as that caused by jetlag and shift work) may have greater physiological costs than non-entrainment (such as that caused by non-24 h sleep/wake disorder). Additionally, it is now understood that bmal1/mice exhibit a higher mortality rate and an expedited onset of age-related diseases beginning at around 26 weeks of age (Kondratov et al. 2006). Similar to humans, clock mutant mice exhibit an accelerated rate of ageing and a 13% reduction in an average lifetime (Dubrovsky et al. 2010) It is uncertain if the disease would develop if the clock and bmal1 mutant mice were allowed to exist in an arrhythmic state continuously, although this subject is intriguing given earlier research on the tau mutant hamster. Research has shown that genetically altering the fly circadian clock by a per null mutation also raises mortality and speeds up ageing (Krishnan et al. 2012).

A loss of circadian rhythmicity is harmful to health; therefore, it makes sense that SCN injuries would shorten lifespan as they cause overt arrhythmia. Even though the majority of investigations have come up empty (e.g., see Dark et al. 1985), it is crucial to remember that most laboratory settings lack the rich biological context that animals require to obtain food and avoid becoming food. It has been claimed that arrhythmic animals with SCN lesions exhibit a lower survival due to an increased predation risk (DeCoursey et al. 1997) or some other unidentified cause in the rare cases where these factors are considered in seminaturalistic conditions (DeCoursey and Krulas 1998). However, recent research on mice with the per2 null mutation kept in semi-naturalistic environments did not discover any detrimental long-term impacts on fitness or survival (Daan et al. 2011).

6.7.2 Cancer and Oxidative Stress

According to recent studies, using artificial light at night increases the risk of breast cancer in sighted women (Kloog et al. 2010) but not in blind women (Hahn 1998). Additionally, shift employment and/or frequent plane travel are linked to an increased risk of developing a number of cancers, including endometrial, colorectal, lymphatic, prostate, breast and other cancers (reviewed in Haus and Smolensky 2013). Additionally, shift workers have epigenetic profiles that point to a higher risk of tumorogenesis (Zhu et al. 2011). Under shift work situations, melatonin suppression and/or disruption of its correct timing may play a significant role in the elevated risk of developing cancer (Haus and Smolensky 2013). Due to these epidemiological data, shift work has been downgraded from a potential to a likely human carcinogen (Wise 2009).

Unlike rats who continue on static LD cycles, nocturnal animals with a fast-rotating simulated jetlag schedule exhibit increased tumour growth. Following a SCN lesion, increased tumour growth is also shown under static LD cycles. Furthermore, continuous light has a negative impact on oxidative stress levels and speeds up tumour growth in animal models (Cos et al. 2006). In rodent models, nighttime lighting is also linked to increased carcinogenesis (Cruz et al. 2003) and oxidative stress variations (Reiter et al. 2000). Given the part that melatonin suppression plays in mediating these effects, it should come as no surprise that adjusting day length, which affects how long melatonin is secreted, can have an impact on tumour growth in mouse models (Nelson and Blom 1994).

The circadian clock controls a wide range of metabolic and cell cycle components, and abnormal circadian gene expression has been seen in a number of malignancies (reviewed in Gery and Koeffler 2007). Per mutant and knockout mice exhibit a significant phenotype in the circadian clock mutant models, including molecular dysregulation of metabolic and cell cycle pathways, increased risk for spontaneous tumour formation, and accelerated growth in induced tumour models (Wood et al. 2008). Furthermore, downregulation of per promotes proliferation whereas overexpression inhibits it in tumour cell cultures (Yang et al. 2009a, b).

The importance of the circadian clock in the appropriate control of the cell cycle and tumour growth is shown by the higher risk for spontaneous and induced tumour growth in bmal1 and cry knockout mouse models (see Gauger and Sancar 2005). Additionally, both bmal1 and cry are involved in controlling oxidative stress; in p53 knockout mice, bmal1 deficiency increases sensitivity and ROS levels (Khapre et al. 2011) whereas cry deficit decreases sensitivity (Ozturk et al. 2009). There have also been reports of links between human clock gene polymorphisms and cancer risk or severity (clock, npas2, per 3, cry1 and timeless; Dai et al. 2011; Monsees et al. 2012; Zhao et al. 2012a, b; Fu et al. 2012).

6.7.3 Metabolic Syndrome, Obesity, and Gastrointestinal Problems

The circadian clock controls metabolic processes and feeding behavior (see Fig. 6.4 for details; Li et al. 2012). Shift workers show a higher incidence of several metabolic syndrome risk factors, including weight gain and obesity, altered free fatty acid, cholesterol, and triglyceride levels, altered lipid and carbohydrate metabolism, insulin resistance, glucose intolerance, and diabetes, as well as changes in appetite-regulating hormones, growth hormones, and c-reactive protein (Wang et al. 2011).

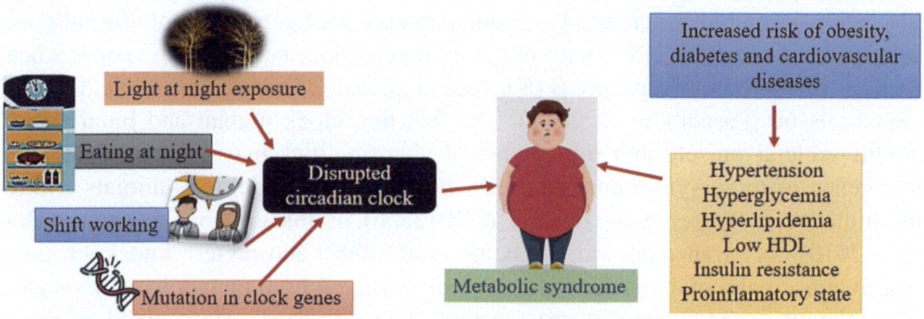

Fig. 6.4 Depiction of how the circadian clock controls metabolic process and feeding behaviour

These risk factors include increased weight gain and obesity, altered free fatty acid, cholesterol, and triglyceride levels, altered. Additionally, shift workers had higher incidences of a number of gastrointestinal conditions, such as gastrointestinal pain, ulcers, and irritable bowel syndrome (Saberi and Moravveji 2010). Additionally, shift workers exhibit different food habits (Pasqua and Moreno 2004), which probably contributes to these illnesses. In humans, simulated shift work has a variety of effects on postprandial hormone levels and metabolic state, depending on the meal's nutritional value. When humans are given isocaloric meals under simulated conditions of circadian misalignment, leptin levels are negatively impacted and glucose levels are badly affected (Scheer et al. 2009). While being sleep-deprived, humans who undergo this form of forced desynchrony procedure experience negative metabolic effects, including a drop in resting metabolism and an increase in postprandial glucose levels (Buxton et al. 2012). In addition, simulated eastward or westward travel in people raised their insulin and glucose levels and boosted the oxidation of carbohydrates while lowering the oxidation of proteins (Gonnissen et al. 2012). Rodent models of simulated shift work have also been linked to metabolic syndrome and obesity (Barclay et al. 2012), which may be related to eating at the wrong times (Bray et al. 2010). Similar to humans, mice exposed to non-24 h LD cycles develop obesity, have faster weight gain, and have higher blood sugar and insulin levels (Karatsoreos et al. 2011). The negative effects of circadian disruption are also seen during development, where repeated changes to the LD cycle predispose rats to sex-dependent metabolic consequences in adulthood, such as increased adiposity, hyperinsulinemia, decreased glucose tolerance, and increased insulin secretion (Varcoe et al. 2011). Additionally, mice kept in continual light exhibit abnormal alterations in their food habits, obesity, and metabolic status (Shi et al. 2013). In addition to changing the time at which food is consumed, increased body mass, and decreased glucose tolerance in mice, the effects of light at night can be mitigated by properly phasing food consumption.

The timing of food intake, body weight gain, adiposity, cholesterol, triglyceride, and insulin levels are altered in mice with a dominant negative mutation in the clock gene. Orexin and ghrelin levels are decreased, and pancreatic function is compromised (Marcheva et al. 2010). Clock/mice also display a metabolic phenotype with hypotension and renal abnormalities. In addition, bmal1/mice exhibit a diabetic phenotype, insulin resistance, ectopic fat formation, elevated levels of circulating fatty acids, impaired pancreatic function, and a reduced genetic profile for adipose function (Shimba et al. 2011) with impaired metabolic function being visible when bmal1 is restricted to the pancreas (Sadacca et al. 2011) liver (Lamia et al. 2008) or adipose tissue (Paschos et al. 2012). The fact that clock mutant and bmal1/mice exhibit skeletal muscle and mitochondrial abnormalities may play a role in the emergence of these symptoms (Andrews et al. 2010). In addition, animals with a null mutation in the per gene (Zhao et al. 2012a, b), double cry knockout mice (Doi et al. 2010) cry1 transgenic mice (Okano et al. 2009) and rev-erb knockout mice exhibit abnormal metabolic function, including altered lipid metabolism, poor glucose tolerance, and/or insulin resistance (Cho et al. 2012). Unresolved in these models is the issue of whether atypical synchronization under 24 h LD cycles or clock disturbances per se cause the metabolic changes.

6.7.4 Cardiovascular Disorders

The peak incidence of cardiac ischemia events, including angina, acute myocardial infarction, and sudden cardiac arrest, is affected by changes in hemodynamic, hemostatic, endothelial, and autonomic factors during the day (reviewed in Chatham and Young 2013). Shift workers exhibit signs of cardiovascular disease, such as atherosclerosis (Knutsson et al. 1999), coronary and ischemic heart disease (Tenkanen et al. 1998), stroke (Brown et al. 2009), and hypertension (Lieu et al. 2012). In people, nighttime light can significantly raise cortisol and heart rate (Cajochen et al. 2005).

Systolic blood pressure, diastolic blood pressure, and heart rate rhythms resynchronize daily at a rate comparable to the rhythm of locomotor activity in rats after the LD cycle is inverted (Campos et al. 2006). Chronic shifting of the LD cycle reduces lifespan, raises blood pressure, and increases expression of cardiovascular risk markers in hypertensive, cardiomyopathic, and obese mouse models (Kishi et al. 2010).

Heterozygote tau-mutant hamsters are shown to exhibit several signs of cardiomyopathy, such as hypotension, myocyte hypertrophy, interstitial fibrosis, and collagen deposition, when kept in 24 h LD cycles that are in contradiction with their natural period length (22 h). Interestingly, these illnesses do not manifest under 22 h LD cycles, indicating that the changed alignment to the external lighting condition and not the genetic mutation itself is to blame for the pathology. Furthermore, the disease was not visible in hamsters with SCN lesions, indicating that a misaligned clock is preferable to none at all. Homozygote taumutant hamsters did not exhibit pathology, possibly as a result of their free-running behaviour during 24-h LD cycles. Recent research has shown that bmal1/mice exhibit cardiomyopathy, endothelial dysfunction, and arteriosclerosis, which provides additional evidence that circadian disturbance is linked to cardiovascular illness (Cheng et al. 2011).

Although bmal1/mice exhibit a number of diseases that might lead to cardiovascular dysfunction, findings from other clock gene models support the idea that the circadian clock itself is necessary for cardiovascular health (Anea et al. 2010; Wang et al. 2008). Similar to the metabolic syndrome, there is some evidence that human clock gene polymorphisms (bmal1229 and clock276) enhance the risk of cardiovascular disease.

6.7.5 Disorders Related to Reproduction

Circadian clocks control reproductive cycles, and reproductive health can be used as a valuable predictor of medical conditions that would make it more difficult to support pregnancy and breastfeeding (Boden et al. 2013). Rotating and night shift workers have a higher chance of developing a variety of reproductive problems than daytime employees, such as irregular menstrual cycles, endometriosis, preterm deliveries, spontaneous abortions, and low birth weight (Mahoney 2010; Lawson et al. 2011).

The timing of behavioural and physiological reproductive events in rodents is regulated by the circadian system, which ensures that it happens at the best times of the day and year (reviewed in Ikegami and Yoshimura 2012). Environmental stimuli that change the circadian phase can change and disrupt the timing of these events (Kennaway et al. 2012). Rats on non-24 h schedules had a smaller luteinizing hormone spike, which is probably caused by a decrease in the phase coherence between the SCN compartments (Smarr et al. 2012). Similarly, under constant light, hamsters with bimodal activity rhythms have two daily spikes in luteinizing hormone with smaller amplitude peaks, likely due to antiphase cycling in the contralateral SCN lobes. Additionally, persistent light in rodents throws off their estrous cycles and prevents South Indian gerbils from reproducing normally (Thomas et al. 2001). The LD cycle's repeated 6 h phase advances, applied every 5 days following copulation, adversely harmed pregnancy outcomes in mice, with normal term deliveries occurring in 11/12 control mice as opposed to only 4/18 advancing animals and 9/18 delaying mice (Summa et al. 2012). Rats exposed to non-24 h LD cycles and constant light (12 h) had lower levels of fertility and preterm foetus weight. In mice kept under non-24 h LD cycles (22 and 26 h), higher levels of embryonic reabsorption and pup mortality after birth are also noticeable (Endo and Watanabe 1989). Environmental circadian disturbance affects reproduction in flies as well, with female flies producing fewer eggs due to decreased food intake during the generally inactive portion of the LD cycle.4.5.2.

Several central and peripheral reproductive organs exhibit clock gene rhythms and appropriate temporal synchronisation within and among tissues is considered crucial for the viability of reproduction. Homozygous clock mutant mice have poor reproductive success, lengthier and erratic estrous cycles, and an undetectable luteinizing hormone surge at the right phase, all of which are consistent with this theory (Dolatshad et al. 2006). When clock mutant mice became arrhythmic after being released into perpetual darkness, irregular estrous cycles were made worse. In addition, clock mutant mice showed increased incidences of pregnancy problems, such as the reabsorption of embryos and fully developed foetuses, as well as challenges giving birth under an LD cycle (Miller et al. 2004) or permanent darkness (Dolatshad et al. 2006). Nevertheless, these effects seem to be strain-dependent, with more severe effects being shown with the clock mutation on the B6 background as opposed to the Balb/c background (Kennaway et al. 2004). The bmal1/ mouse model has also produced some results that are contradictory. Initial findings did not mention any reproductive issues, but recent investigations have revealed that bmal1/mice have erratic estrous cycles and poor fecundity (Ratajczak et al. 2009). Despite having fecundity comparable to wild-type mice at 2–6 months of age, female mice lacking per1 or per2 also exhibit irregular estrous cycles and lower fertility towards middle age (9–12 months; Pilorz and Steinlechner 2008). While there hasn't been much research on potential links between human clock gene polymorphisms and reproductive problems, one study found a connection between bmal1 and npas2 variations and miscarriage (Kovanen et al. 2010). The studied pathways of clock's regulation of reproduction are depicted in Fig. 6.5.

6.7.6 Mood Disorders

The fact that many psychological disorders are characterized by sleep and circadian disruptions and that both physiological and circadian symptoms can be reduced with timed administration of bright light and/or melatonin (reviewed in Lanfumey et al. 2013) suggest that there is a close relationship between circadian regulation and mood. While jet lag and shift work may not always result in psychotic or serious affective disorders, they can cause a return of symptoms in those who have previously had psychiatric disease (Katz et al. 2002). Additionally, the severity of symptoms in nonseasonal depression is correlated with the chronotype and amount of misalignment in individuals (Emens et al. 2009) and the winter blues (Lewy et al. 2009).

Experimental research has focused on the nature of the connection between seasonal depression and variations in circadian function brought on by changes in light exposure. Depression is linked to entrainment to short photoperiods and reduced light exposure in diurnal rodents, and it may also be seen in some nocturnal rat species (Krivisky et al. 2011). However, the species, strain, behavioural test used, and time of day assessed can all have a significant impact on whether or not nocturnal rats' affective responses are affected by day duration (Flaisher-Grinberg et al. 2011). In nocturnal rodents, increased light exposure is also linked to altered affective

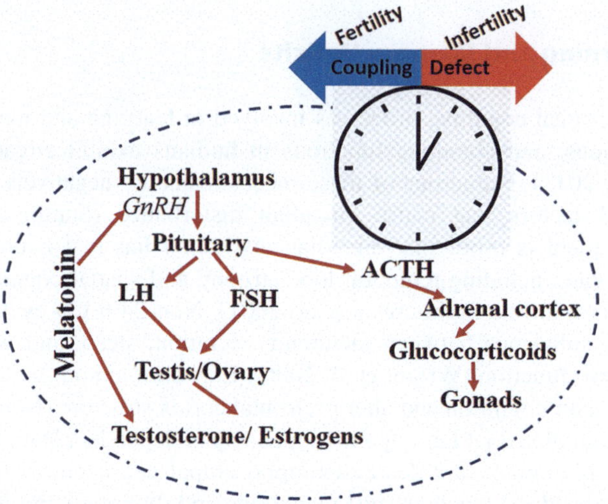

Fig. 6.5 Schematic representation of pathways by which clock influence reproduction in animals

states; mice kept in continual light showed behavioural indicators of depression and reduced anxiety (Fonken et al. 2010).

Furthermore, both entrained (Bedrosian et al. 2011) and free-running mice exposed to light at night display anxious- and depressive-like behaviour. Depression-like behaviour can be improved by administering fluoxetine or by ablating retinal ganglion cells that are innately photosensitive and contain melanopsin in the presence of aberrant light at night. Lastly, when evaluated in novel contextual settings, mice exposed to 20-h LD cycles show an elevated impulsivity phenotype. In contrast, mice subjected to a weekly 6 h LD shift did not exhibit anxious or depressive behavior (Castanon-Cervantes et al. 2010).

These local clocks may help to regulate mood because clock gene expression rhythms are seen in a number of structures crucial for motivation, reward processing, and mood regulation (Guilding and Piggins 2007). Studies using the clock mutant mouse, which has been suggested as a model for mania due to its hyperactivity, decreased depressive- and anxiety-like behaviour, and increased reward- and goal-directed behaviour, have shown the strongest correlation between clock gene function and mood to date (Easton et al. 2003; Roybal et al. 2007). The fact that lithium therapy or clock rescue in the ventral tegmental region can improve this behavioural profile adds to the credibility of the clock mutant concept for mania (Roybal et al. 2007). Interestingly, RNAi disruption of clock in the ventral tegmental area can cause both manic- and depressive-like behaviours, highlighting the significance of clock gene expression in this particular brain location for healthy mood regulation (Mukherjee et al. 2010). The ROR knockout mouse, the per2 mutant mouse, and the fbxl3 mutant mouse, which has a 27-h period because of altered cry degradation, are all reported to have less despair and anxiety (Keers et al. 2012). In addition, a number of human genetic association studies (reviewed in McCarthy et al. 2012) have linked clock genes to the aetiology of mood disorders.

6.7.7 Learning and Memory Deficits

Many of the critical cognitive processes involved in learning and memory exhibit daily oscillations, and cognitive functions in humans exhibit circadian changes (Wright et al. 2012). Numerous of these mechanisms are negatively impacted by environmental factors that cause circadian disturbance (Santhi et al. 2007). Additionally, there is proof that circadian disruption has real neuropathological effects on people, including temporal lobe atrophy and spatial cognition deficiencies in long-term airline flight crews (Cho 2001). Non-24 h LD cycles in humans that cause an imbalance between melatonin secretion, sleep, and waking cycles impair cognitive function (Wright et al. 2006). Exposure to non-24 h LD cycles in rodents can impair cognition and alter prefrontal cortex structure (Neto et al. 2008). Furthermore, simulated jet lag impairs hippocampus-dependent learning and memory functions in animals and decreases hippocampal neurogenesis (Gibson et al. 2010). Both procedural (such as shift magnitude and direction) and intrinsic (such as the history of jetlag) factors can have an impact on the deficits brought on by jetlag and whether they remain after adjusting to the shifted LD cycle. Rats kept in a bright environment have learning deficits linked to alterations in hippocampus function (Ma et al. 2007). These light-induced learning losses in the latter situation depend on the presence of retinal ganglion cells that are naturally photosensitive and contain melanopsin. In a delayed task requiring novel object recognition, arrhythmic Siberian hamsters exhibit learning difficulties (Ruby et al. 2008). Numerous learning-related structures exhibit clock gene expression rhythms, and these local clocks may support memory functions (Wang et al. 2009).

Studies showing that flies with a per null mutation demonstrated learning problems in an experience-dependent courting paradigm provided the first evidence that clock genes may play a role in learning processes (Sakai et al. 2004). Building on this initial research, it has been shown that mice with a per2 null mutation show deficiencies in trace fear conditioning and long-term potentiation, with normal short-term recall and initial acquisition. Similarly, learning problems in a radial arm maze task are seen in per1 knockout animals, who also have attenuated hippocampus clock gene rhythms (Jilg et al. 2010). Additionally, mice lacking either the per1 or per2 gene exhibit altered cocaine-induced behavioural sensitization and location preference, with the per1 mutant and the per2 mutant exhibiting, respectively, diminished and heightened responses (Abarca et al. 2002). However, mice lacking either of these genes do not exhibit abnormalities in cued fear conditioning, spatial learning in the Morris Water maze, or ethanol-induced learning, raising questions about the relevance of per for learning in general (Zghoul et al. 2007). However, mice missing the clock paralog npas2 do not exhibit deficits in the Morris water maze or cued fear conditioning (Garcia et al. 2000). While clock mutant mice, double cry/mice, and bmal1/mice exhibit different behaviours in response to a habituation protocol, these models need be investigated in the context of additional learning paradigms (Crawley and Paylor 1997).

6.7.8 Immune Dysfunction

Immune response and function are strongly influenced by circadian regulation (reviewed in Scheiermann et al. 2013). Rotating and night shift workers are more likely to contract common infections, multiple sclerosis, and other autoimmune diseases than daytime employees (Magrini et al. 2006). A number of the illnesses seen under circadian disturbance, including cardiovascular disease, cancer, and diabetes, are at risk due to uncontrolled or chronic inflammation (Goldberg 2009). Therefore, immune system modifications may be a common component impacting the various negative health effects related to shift employment. Using a subcutaneous injection with peripheral blood stimulation in a popliteal lymph node assay, chronic LD cycle reversals have been shown to impair immunological function in rats, evident by the reduced cellular responses to concanavalin (Kort and Weijma 1982). On the other hand, it was discovered that long-term reversals of the LD cycle in mice sensitized the inflammatory response to the colitis model dextran sodium sulphate. The immunological response to lipopolysaccharide, a model for sepsis, was worse by weekly

6 h changes in the LD cycle in mice. Repeated LD shifts also reduced natural killer cell cytotoxicity and aided in the growth of lung tumours. Similarly, immune system performance is impacted by persistent light exposure in rodent and bird species (Moore and Siopes 2000).

Siberian hamsters' immune systems are also inhibited by exposure to light at night. Chronic exposure results in decreased delayed-type hypersensitivity and diminished blood plasma bactericidal activity after the LPS challenge. In the Siberian hamster, seasonal variations in immunological responses have been observed (Prendergast 2008). Additionally, rats, mice, and flies develop immunological dysfunction as a result of frequent and abrupt changes in day length.

Different types of immune cells exhibit clock gene expression rhythms, and local clock activity is crucial for the temporal control of inflammatory responses (see Fig. 6.6; Scheiermann et al. 2013). With rev-erba/mice exhibiting a hypersensitive response to the lipopolysaccharide challenge, rev-erba has been identified as a crucial component of the system by which the circadian clock regulates the temporal gating of immune activity (Gibbs et al. 2012). Additionally, it has been discovered that the clock gene CRY adversely controls cytokine synthesis (Narasimamurthy et al. 2012) and mice who have both of the cry paralogs deleted exhibit an immunological phenotype that includes increased cytokine production and joint swelling following arthritis induction. Additionally, Natural Killer cells from per2 mutant mice and per1/animals exhibit altered cytokine and cytolytic expression patterns 412 and per2 mutant mice exhibit decreased responses to the LPS challenge (Liu et al. 2006). Mice lacking bmal1 have increasing ocular irritation and reduced immune cell expression (Sun et al. 2006). Mice with a clock null mutation have altered immune gene transcription (Oishi et al. 2003) and diminished NF-kB

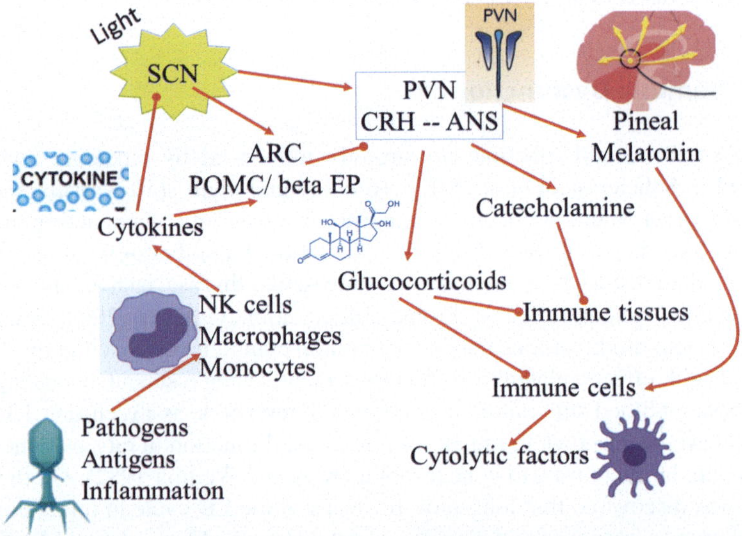

Fig. 6.6 Representation of clock activity on the temporal control of inflammatory responses

activation in cultured fibroblasts and hepatocytes (Spengler et al. 2012). A correlation between a particular per3 polymorphism and the prevalence or severity of inflammatory bowel disease has been found in genetic association studies looking at the interaction between clock gene polymorphisms and immunological illnesses (Mazzoccoli et al. 2012).

6.8 Future Research Questions

Overall, the epidemiological and experimental data amassed up to this point make a strong case that circadian disturbance brought on by environmental and genetic factors has a wide range of negative health effects. Although genetic variants that change clock function may potentially have an impact on health, the environment is the main cause of circadian clock disruption in humans. The present 24 h lifestyle is made possible and encouraged by artificial light and rising globalization, and unconventional work schedules are becoming more and more frequent. While there has been research into the elements that affect employee productivity, there has been much less investigation into the potential short- and long-term health effects of staffing techniques that are intended to provide coverage throughout the day and night. Therefore, a deeper comprehension of the impacts of unconventional schedules and the ways in which these circumstances have an impact on health is required. Shift employment frequently causes the circadian clock to be out of sync with external time, inadequate sleep, suppression of the melatonin hormone, psychological stress, and adjustments to other behavioural health variables such as eating habits and alcohol and nicotine use. Animal models offer the chance to separate these factors to determine the relative effects of each on particular health outcomes. For instance, increasing light exposure in mice can induce sleep (Altimus et al. 2008) rather than result in sleep deprivation, allowing the results to differentiate between mechanisms associated with the circadian clock and those related to sleep. Similarly, most mouse breeds exhibit functioning receptors for the hormone melatonin (Ebihara et al. 1986) but lack the enzymes necessary for its synthesis. Furthermore, although this may depend on the schedule, mice can be subjected to schedules interrupting the circadian rhythm without triggering a stress response or altering their anxious or depressive behavior. Therefore, the nocturnal mouse offers a suitable model to investigate the role of circadian disturbance further. The extent of the health effects associated with various work and rest patterns is still a significant open subject. Since it has been challenging to distinguish between various shift work paradigms in epidemiological studies up to this point, it is still unclear how different shift work schedules affect sleep, cognition, and health-related parameters (Sallinen and Kecklund 2010). Fixed or rotating shifts are both possible, with rotating shifts moving either forward (delaying) or backward (advancing). Studies using animal models have shown that shift sizes of 6 h had a detrimental effect. The few studies that have addressed these kinds of issues imply that schedule factors like shift orientation and exposure length considerably influence how detrimental circadian disturbance is to one's health. Shift work scheduling would benefit from more research evaluating the effects of various schedules on particular health metrics in both human and animal models.

Additional studies exploring the relevance of exposure duration across schedule types, the durability of health effects, and individual susceptibility variables would be of the most value. The role of central vs. local clock function in the consequences of circadian disruption is still a topic of investigation. All cells in humans should be affected by clock gene polymorphisms. Most genetic models used up to this point exhibit similar widespread impacts on cells in both central and peripheral clocks. The role of local clock function in determining health outcomes has been supported by various papers that have reported the health effects of tissue-specific genetic modifications over the past few years. The development of SCN-specific gene targeting strategies is still in progress. However, environmental models that interfere with SCN activity and system-level regulation offer useful models and may be quite reflective of the particular circumstances that frequently cause circadian disturbance in humans. It is unknown to what extent changes in central or peripheral function (such as a loss of synchronization within or among peripheral clocks) result in health implications when environmental lighting is altered. Using tissue-specific profiling during disease states and independent modification of implanted cell and host genomes during tumour formation, cancer research is one area where these questions have started to be addressed (Zeng et al. 2010). Research investigating the relative effects of losing normal circadian timing in central versus peripheral regions can be facilitated by recent developments in cell-specific in vivo expression technology.

The role of a malfunctioning clock as opposed to a broken clock in illness genesis is the last area that requires more understanding. Excellent tools are provided here by both environmental and nonarrhythmic genetic models, but work to date has occasionally fallen short of fully describing the rhythmic state of the model at the large (e.g., behavioural and hormonal) and fine (cellular) levels of study. The rhythmic state of the animals should be documented in experiments that use steady light, for instance, giving researchers a chance to link health consequences to the severity and timing of animal circadian disruption. Furthermore, it is yet unknown how differing entrainment and free-running settings (such as LD vs. DD conditions) affect illness states in arrhythmic clock gene models (such as bmal1/, clock mutant, double per/, or cry/). Finally, rhythmic clock gene models (such as the per2 mutant) maintained under both 24 h LD cycles and LD cycles with a period closer to the inherent period of the mouse model can be used to study the effects of circadian disruption on health in more detail. By comparing the effects of misalignment and arrhythmia, research can assess whether having the appropriate clock for the environment is corrective and whether there are any circumstances in which a malfunctioning clock can be useful.

6.9 Conclusion

There is strong evidence that disrupting circadian rhythms is bad for health, and the role of circadian timing in disease manifestation is becoming more widely recognized. Most of the studies in this review suggest that the organism can correct circadian disturbance when the environment is free of challenge or stress; nevertheless, serious health effects occur when circadian control is weakened in the presence of difficulty. All of these would seem to indicate that while circadian disturbance may not directly cause disease, it can interact with disease states to worsen outcomes and/or hasten the progression of disease. Future research should focus on understanding the molecular connections that underlie these synergistic and modulatory impacts of circadian disturbance on health.

References

Abarca C, Albrecht U, Spanagel R (2002) Cocaine sensitization and reward are under the influence of circadian genes and rhythm. Proc Natl Acad Sci U S A 99:9026–9030

Akerstedt T, Wright KP Jr (2009) Sleep loss and fatigue in shift work and shift work disorder. Sleep Med Clin 4:257–271

Altimus CM, Guler AD, Villa KL et al (2008) Rods-cones and melanopsin detect light and dark to modulate sleep independent of image formation. Proc Natl Acad Sci U S A 105:19998–20003

Andrews JL, Zhang X, McCarthy JJ et al (2010) CLOCK and BMAL1 regulate MyoD and are necessary for maintenance of skeletal muscle phenotype and function. Proc Natl Acad Sci U S A 107:19090–19095

Anea CB, Ali MI, Osmond JM et al (2010) Matrix metalloproteinase 2 and 9 dysfunction underlie vascular stiffness in circadian clock mutant mice. Arterioscler Thromb Vasc Biol 30:2535–2543

Arble DM, Ramsey KM, Bass J, Turek FW (2010) Circadian disruption and metabolic disease: findings from animal models. Best Pract Res Clin Endocrinol Metab 24(5):785–800. https://doi.org/10.1016/j.beem.2010.08.003

Aschoff J, von Saint PU, Wever R (1971) Lifetime of flies under influence of time displacement. Naturwissenschaften 58:574

Barclay JL, Husse J, Bode B et al (2012) Circadian desynchrony promotes metabolic disruption in a mouse model of shiftwork. PLoS One 7:e37150

Bedrosian TA, Fonken LK, Walton JC, Nelson RJ (2011) Chronic exposure to dim light at night suppresses immune responses in Siberian hamsters. Biol Lett 7:468–471

Bedrosian TA, Fonken LK, Walton JC, Haim A, Nelson RJ (2011) Dim light at night provokes depression-like behaviors and reduces CA1 dendritic spine density in female hamsters. Psychoneuroendocrinology 36:1062–1069

Boden MJ, Varcoe TJ, Kennaway DJ (2013) Circadian regulation of reproduction: from gamete to offspring. Prog Biophys Mol Biol 113:387. https://doi.org/10.1016/j.pbiomolbio.2013.01.003

Bray MS, Tsai JY, Villegas-Montoya C et al (2010) Time-of-day-dependent dietary fat consumption influences multiple cardiometabolic syndrome parameters in mice. Int J Obes 34:1589–1598

Brown DL, Feskanich D, Sanchez BN et al (2009) Rotating night shift work and the risk of ischemic stroke. Am J Epidemiol 169:1370–1377

Buhr ED, Takahashi JS (2013) Molecular components of the mammalian circadian clock. Handb Exp Pharmacol 217:3–27. https://doi.org/10.1007/978-3-642-25950-0_1

Bunger MK, Wilsbacher LD, Moran SM et al (2000) Mop3 is an essential component of the master circadian pacemaker in mammals. Cell 103:1009–1017

Buxton OM, Cain SW, O'Connor SP et al (2012) Adverse metabolic consequences in humans of prolonged sleep restriction combined with circadian disruption. Sci Transl Med 4:129ra143

Cajochen C, Munch M, Kobialka S et al (2005) High sensitivity of human melatonin, alertness, thermoregulation, and heart rate to short wavelength light. J Clin Endocrinol Metab 90:1311–1316

Cambras T, Weller JR, Anglès-Pujoràs M, Lee ML, Christopher A, Díez-Noguera A, Krueger JM, De La Iglesia HO (2007) Circadian desynchronization of core body temperature and sleep stages in the rat. Proc Natl Acad Sci 104(18):7634–7639. https://doi.org/10.1073/pnas.0702424104

Campos LA, Plehm R, Cipolla-Neto J, Bader M, Baltatu OC (2006) Altered circadian rhythm reentrainment to light phase shifts in rats with low levels of brain angiotensinogen. Am J Physiol Regul Integr Comp Physiol 290:R1122–R1127

Castanon-Cervantes O, Wu M, Ehlen JC et al (2010) Dysregulation of inflammatory responses by chronic circadian disruption. J Immunol 185:5796–5805

Chatham JC, Young ME (2013) Regulation of myocardial metabolism by the cardiomyocyte circadian clock. J Mol Cell Cardiol 55:139–146

Cheng B, Anea CB, Yao L et al (2011) Tissue-intrinsic dysfunction of circadian clock confers transplant arteriosclerosis. Proc Natl Acad Sci U S A 108:17147–17152

Chew KS, Renna JM, McNeill DS, Fernandez DC, Keenan WT, Thomsen MB, Ecker JL, Loevinsohn GS, VanDunk C, Vicarel DC, Tufford A, Weng S, Gray PA, Cayouette M, Herzog ED, Zhao H, Berson DM, Hattar S (2017) A subset of ipRGCs regulates both maturation of the circadian clock and segregation of retinogeniculate projections in mice. eLife 6:e22861. https://doi.org/10.7554/eLife.22861

Cho K (2001) Chronic 'jet lag' produces temporal lobe atrophy and spatial cognitive deficits. Nat Neurosci 4:567–568

Cho H, Zhao X, Hatori M et al (2012) Regulation of circadian behaviour and metabolism by REV-ERB-alpha and REV-ERB-beta. Nature 485:123–127

Cos S, Mediavilla D, Martinez-Campa C et al (2006) Exposure to light-at-night increases the growth of DMBA-induced mammary adenocarcinomas in rats. Cancer Lett 235:266–271

Costa G (1997) The problem: shiftwork. Chronobiol Int 14:89–98

Crawley JN, Paylor R (1997) A proposed test battery and constellations of specific behavioural paradigms to investigate the behavioral phenotypes of transgenic and knockout mice. Horm Behav 31:197–211

Cruz A, Padillo FJ, Granados J et al (2003) Effect of melatonin on cholestatic oxidative stress under constant light exposure. Cell Biochem Funct 21:377–380

Daan S, Spoelstra K, Albrecht U et al (2011) Lab mice in the field: unorthodox daily activity and effects of a dysfunctional circadian clock allele. J Biol Rhythm 26:118–129

Dai H, Zhang L, Cao M et al (2011) The role of polymorphisms in circadian pathway genes in breast tumorigenesis. Breast Cancer Res Treat 127:531–540

Dark J, Pickard GE, Zucker I (1985) Persistence of circannual rhythms in ground squirrels with lesions of the suprachiasmatic nuclei. Brain Res 332:201–207

Davidson AJ, Sellix MT, Daniel J et al (2006) Chronic jetlag increases mortality in aged mice. Curr Biol 16:R914–R916

de la Iglesia HO, Meyer J, Carpino A Jr, Schwartz WJ (2000) Antiphase oscillation of the left and right suprachiasmatic nuclei. Science 290:799–801

de la Iglesia HO, Cambras T, Schwartz WJ, Diez-Noguera A (2004) Forced desynchronization of dual circadian oscillators within the rat suprachiasmatic nucleus. Curr Biol 14:796–800

DeCoursey PJ, Krulas JR (1998) Behavior of SCN-lesioned chipmunks in natural habitat: a pilot study. J Biol Rhythm 13:229–244

DeCoursey PJ, Krulas JR, Mele G, Holley DC (1997) Circadian performance of suprachiasmatic nuclei (SCN)-lesioned antelope ground squirrels in a desert enclosure. Physiol Behav 62:1099–1108

Dijk DJ, Duffy JF, Riel E, Shanahan TL, Czeisler CA (1999) Ageing and the circadian and homeostatic regulation of human sleep during forced desynchrony of rest, melatonin and temperature rhythms. J Physiol 516:611–627

Doi M, Takahashi Y, Komatsu R et al (2010) Salt-sensitive hypertension in circadian clock deficient Cry-null mice involves dysregulated adrenal Hsd3b6. Nat Med 16:67–74

Dolatshad H, Campbell EA, O'Hara L et al (2006) Developmental and reproductive performance in circadian mutant mice. Hum Reprod 21:68–79

Dubrovsky YV, Samsa WE, Kondratov RV (2010) Deficiency of circadian protein CLOCK reduces lifespan and increases age-related cataract development in mice. Aging 2:936–944

Dumont M, Lanctot V, Cadieux-Viau R, Paquet J (2012) Melatonin production and light exposure of rotating night workers. Chronobiol Int 29:203–210

Easton A, Arbuzova J, Turek FW (2003) The circadian Clock mutation increases exploratory activity and escapeseeking behavior. Genes Brain Behav 2:11–19

Ebihara S, Marks T, Hudson DJ, Menaker M (1986) Genetic control of melatonin synthesis in the pineal gland of the mouse. Science 231:491–493

Emens J, Lewy A, Kinzie JM, Arntz D, Rough J (2009) Circadian misalignment in major depressive disorder. Psychiatry Res 168:259–261

Endo A, Watanabe T (1989) Effects of non-24-hour days on reproductive efficacy and embryonic development in mice. Gamete Res 22:435–441

Erren TC, Reiter RJ, Piekarski C (2003) Light, timing of biological rhythms, and chronodisruption in man. Naturwissenschaften 90(11):485–494. https://doi.org/10.1007/s00114-003-0468-6

Fagiani F, Baronchelli E, Pittaluga A, Pedrini E, Scacchi C, Govoni S, Lanni C (2022) The circadian molecular machinery in CNS cells: a fine tuner of neuronal and glial activity with space/time resolution. Front Mol Neurosci 15:937174. https://doi.org/10.3389/fnmol.2022.937174

Flaisher-Grinberg S, Gampetro DR, Kronfeld-Schor N, Einat H (2011) Inconsistent effects of photoperiod manipulations in tests for affective-like changes in mice: implications for the selection of appropriate model animals. Behav Pharmacol 22:23–30

Folkard S, Tucker P (2003) Shift work, safety and productivity. Occup Med 53:95–101

Fonken LK, Workman JL, Walton JC et al (2010) Light at night increases body mass by shifting the time of food intake. Proc Natl Acad Sci U S A 107:18664–18669

Fu A, Leaderer D, Zheng T et al (2012) Genetic and epigenetic associations of circadian gene TIMELESS and breast cancer risk. Mol Carcinog 51:923–929

Garcia JA, Zhang D, Estill SJ et al (2000) Impaired cued and contextual memory in NPAS2-deficient mice. Science 288:2226–2230

Gauger MA, Sancar A (2005) Cryptochrome, circadian cycle, cell cycle checkpoints, and cancer. Cancer Res 65:6828–6834

Gery S, Koeffler HP (2007) The role of circadian regulation in cancer. Cold Spring Harb Symp Quant Biol 72:459–464

Gibbs JE, Blaikley J, Beesley S et al (2012) The nuclear receptor REV-ERBalpha mediates circadian regulation of innate immunity through selective regulation of inflammatory cytokines. Proc Natl Acad Sci U S A 109:582–587

Gibson EM, Wang C, Tjho S, Khattar N, Kriegsfeld LJ (2010) Experimental 'jet lag' inhibits adult neurogenesis and produces long-term cognitive deficits in female hamsters. PLoS One 5:e15267

Goldberg RB (2009) Cytokine and cytokine-like inflammation markers, endothelial dysfunction, and imbalanced coagulation in development of diabetes and its complications. J Clin Endocrinol Metab 94:3171–3182

Gonnissen HK, Rutters F, Mazuy C et al (2012) Effect of a phase advance and phase delay of the 24-h cycle on energy metabolism, appetite, and related hormones. Am J Clin Nutr 96:689–697

Gronfier C, Wright KP Jr, Kronauer RE, Czeisler CA (2007) Entrainment of the human circadian pacemaker to longer-than-24-h days. Proc Natl Acad Sci U S A 104:9081–9086

Guilding C, Piggins HD (2007) Challenging the omnipotence of the suprachiasmatic timekeeper: are circadian oscillators present throughout the mammalian brain? Eur J Neurosci 25:3195–3216

Hahn RA (1998) Profound bilateral blindness and the incidence of breast cancer. Epidemiology 2:208–210

Halberg F, Cadotte I (1975) Increased mortality in mice exposed to weekly 180 shifts of lighting regimine LD 12:12 beginning at 1 year of age. Chronobiologia 2:26

Hattar S, Kumar M, Park A, Tong P, Tung J, Yau K-W, Berson DM (2006) Central projections of melanopsin-expressing retinal ganglion cells in the mouse. J Comp Neurol 497(3):326–349. https://doi.org/10.1002/cne.20970

Haus EL, Smolensky MH (2013) Shift work and cancer risk: potential mechanistic roles of circadian disruption, light at night, and sleep deprivation. Sleep Med Rev 17:273–284

Honma S, Honma K (1999) Light-induced uncoupling of multioscillatory circadian system in a diurnal rodent, Asian chipmunk. Am J Physiol 276:R1390–R1396

Hurd MW, Ralph MR (1998) The significance of circadian organization for longevity in the golden hamster. J Biol Rhythm 13:430–436

Ikegami K, Yoshimura T (2012) Circadian clocks and the measurement of daylength in seasonal reproduction. Mol Cell Endocrinol 349:76–81

Jagannath A, Taylor L, Wakaf Z, Vasudevan SR, Foster RG (2017) The genetics of circadian rhythms, sleep and health. Hum Mol Genet 26(R2):R128–R138

Jewett ME, Kronauer RE, Czeisler CA (1991) Light-induced suppression of endogenous circadian amplitude in humans. Nature 350:59–62

Jewett ME, Forger DB, Kronauer RE (1999) Revised limit cycle oscillator model of human circadian pacemaker. J Biol Rhythm 14:493–499

Jilg A, Lesny S, Peruzki N et al (2010) Temporal dynamics of mouse hippocampal clock gene expression support memory processing. Hippocampus 20:377–388

Karatsoreos IN, Bhagat S, Bloss EB, Morrison JH, McEwen BS (2011) Disruption of circadian clocks has ramifications for metabolism, brain, and behavior. Proc Natl Acad Sci U S A 108:1657–1662

Karlsson B, Alfredsson L, Knutsson A, Andersson E, Toren K (2005) Total mortality and cause-specific mortality of Swedish shift- and dayworkers in the pulp and paper industry in 1952-2001. Scand J Work Environ Health 31:30–35

Katz G, Knobler HY, Laibel Z, Strauss Z, Durst R (2002) Time zone change and major psychiatric morbidity: the results of a 6-year study in Jerusalem. Compr Psychiatry 43:37–40

Keers R, Pedroso I, Breen G et al (2012) Reduced anxiety and depression-like behaviours in the circadian period mutant mouse afterhours. PLoS One 7:e38263

Kelly TL, Neri DF, Grill JT et al (1999) Nonentrained circadian rhythms of melatonin in submariners scheduled to an 18-hour day. J Biol Rhythm 14:190–196

Kennaway DJ, Boden MJ, Voultsios A (2004) Reproductive performance in female Clock Delta19 mutant mice. Reprod Fertil Dev 16:801–810

Kennaway DJ, Boden MJ, Varcoe TJ (2012) Circadian rhythms and fertility. Mol Cell Endocrinol 349:56–61

Kerin A, Aguirre A (2004) Improving health, safety, and profits in extended hours operations (shiftwork). Ind Health 2005(3):201–208

Khapre RV, Kondratova AA, Susova O, Kondratov RV (2011) Circadian clock protein BMAL1 regulates cellular senescence in vivo. Cell Cycle 10:4162–4169

Kishi T, Yoshimura R, Kitajima T, Okochi T, Okumura T, Tsunoka T, Yamanouchi Y, Kinoshita Y, Kawashima K, Fukuo Y, Naitoh H (2010) SIRT1 gene is associated with major depressive disorder in the Japanese population. J Affect Disord 126(1-2):167–173

Kloog I, Stevens RG, Haim A, Portnov BA (2010) Nighttime light level co-distributes with breast cancer incidence worldwide. Cancer Causes Control 21:2059–2068

Knutsson A, Hallquist J, Reuterwall C, Theorell T, Akerstedt T (1999) Shiftwork and myocardial infarction: a case-control study. Occup Environ Med 56:46–50

Kondratov RV, Kondratova AA, Gorbacheva VY, Vykhovanets OV, Antoch MP (2006) Early aging and age-related pathologies in mice deficient in BMAL1, the core component of the circadian clock. Genes Dev 20:1868–1873

Kort WJ, Weijma JM (1982) Effect of chronic light-dark shift stress on the immune response of the rat. Physiol Behav 29:1083–1087

Kovanen L, Saarikoski ST, Aromaa A, Lonnqvist J, Partonen T (2010) ARNTL (BMAL1) and NPAS2 gene variants contribute to fertility and seasonality. PLoS One 5:e10007

Krishnan N, Rakshit K, Chow ES et al (2012) Loss of circadian clock accelerates aging in neurodegeneration-prone mutants. Neurobiol Dis 45:1129–1135

Krivisky K, Ashkenazy T, Kronfeld-Schor N, Einat H (2011) Antidepressants reverse short photoperiod-induced, forced swim test depression-like behavior in the diurnal fat sand rat: further support for the utilization of diurnal rodents for modeling affective disorders. Neuropsychobiology 63:191–196

Lamia KA, Storch KF, Weitz CJ (2008) Physiological significance of a peripheral tissue circadian clock. Proc Natl Acad Sci USA 105:15172–15177

Lanfumey L, Mongeau R, Hamon M (2013) Biological rhythms and melatonin in mood disorders and their treatments. Pharmacol Ther 138:176–184

Lawson CC, Whelan EA, Lividoti Hibert EN et al (2011) Rotating shift work and menstrual cycle characteristics. Epidemiology 22:305–312

LeGates TA, Altimus CM, Wang H et al (2012) Aberrant light directly impairs mood and learning through melanopsin-expressing neurons. Nature 491:594–598

Lewy AJ, Emens JS, Songer JB et al (2009) Winter depression: integrating mood, circadian rhythms, and the sleep/wake and light/dark cycles into a bio-psycho-social environmental model. Sleep Med Clin 4:285–299

Li MD, Li CM, Wang Z (2012) The role of circadian clocks in metabolic disease. Yale J Biol Med 85:387–401

Lieu SJ, Curhan GC, Schernhammer ES, Forman JP (2012) Rotating night shift work and disparate hypertension risk in African-Americans. J Hypertens 30:61–66

Liu J, Malkani G, Shi X et al (2006) The circadian clock Period 2 gene regulates gamma interferon production of NK cells in host response to lipopolysaccharide-induced endotoxic shock. Infect Immun 74:4750–4756

Lockley SW, Dijk DJ, Kosti O, Skene DJ, Arendt J (2008) Alertness, mood and performance rhythm disturbances associated with circadian sleep disorders in the blind. J Sleep Res 17:207–216

Lowrey PL, Shimomura K, Antoch MP et al (2000) Positional syntenic cloning and functional characterization of the mammalian circadian mutation tau. Science 288:483–492

Ma WP, Cao J, Tian M et al (2007) Exposure to chronic constant light impairs spatial memory and influences longterm depression in rats. Neurosci Res 59:224–230

Magrini A, Pietroiusti A, Coppeta L et al (2006) Shift work and autoimmune thyroid disorders. Int J Immunopathol Pharmacol 19:31–36

Mahoney MM (2010) Shift work, jet lag, and female reproduction. Int J Endocrinol 2010:813764

Marcheva B, Ramsey KM, Buhr ED et al (2010) Disruption of the clock components CLOCK and BMAL1 leads to hypoinsulinaemia and diabetes. Nature 466:627–631

Mazzoccoli G, Palmieri O, Corritore G et al (2012) Association study of a polymorphism in clock gene PERIOD3 and risk of inflammatory bowel disease. Chronobiol Int 29:994–1003

Miller BH, Olson SL, Turek FW et al (2004) Circadian clock mutation disrupts estrous cyclicity and maintenance of pregnancy. Curr Biol 14:1367–1373

Monsees GM, Kraft P, Hankinson SE, Hunter DJ, Schernhammer ES (2012) Circadian genes and breast cancer susceptibility in rotating shift workers. Int J Cancer 131:2547–2552

Moore CB, Siopes TD (2000) Effects of lighting conditions and melatonin supplementation on the cellular and humoral immune responses in Japanese quail Coturnix coturnix japonica. Gen Comp Endocrinol 119:95–104

Mukherjee S, Coque L, Cao JL et al (2010) Knockdown of Clock in the ventral tegmental area through RNA interference results in a mixed state of mania and depression-like behavior. Biol Psychiatry 68:503–511

Narasimamurthy R, Hatori M, Nayak SK et al (2012) Circadian clock protein cryptochrome regulates the expression of proinflammatory cytokines. Proc Natl Acad Sci U S A 109:12662–12667

Natelson BH, Ottenweller JE, Tapp WN, Bergen M, Soldan S (1996) Phototherapeutic effects in hamsters with heart disease. Physiol Behav 60:463–468

Nelson RJ, Blom JM (1994) Photoperiodic effects on tumor development and immune function. J Biol Rhythm 9:233–249

Neto SP, Carneiro BT, Valentinuzzi VS, Araujo JF (2008) Dissociation of the circadian rhythm of locomotor activity in a 22 h light-dark cycle impairs passive avoidance but not object recognition memory in rats. Physiol Behav 94:523–527

Oishi K, Miyazaki K, Kadota K et al (2003) Genome-wide expression analysis of mouse liver reveals CLOCK-regulated circadian output genes. J Biol Chem 278:41519–41527

Okano S, Akashi M, Hayasaka K, Nakajima O (2009) Unusual circadian locomotor activity and pathophysiology in mutant CRY1 transgenic mice. Neurosci Lett 451:246–251

Ozturk N, Lee JH, Gaddameedhi S, Sancar A (2009) Loss of cryptochrome reduces cancer risk in p53 mutant mice. Proc Natl Acad Sci U S A 106:2841–2846

Paschos GK, Ibrahim S, Song WL et al (2012) Obesity in mice with adipocyte-specific deletion of clock component Arntl. Nat Med 18:1768–1777

Pasqua IC, Moreno CR (2004) The nutritional status and eating habits of shift workers: a chronobiological approach. Chronobiol Int 21:949–960

Penev PD, Kolker DE, Zee PC, Turek FW (1998) Chronic circadian desynchronization decreases the survival of animals with cardiomyopathic heart disease. Am J Phys 275:H2334–H2337

Perret M (1997) Change in photoperiodic cycle affects life span in a prosimian primate (Microcebus murinus). J Biol Rhythm 12:136–145

Pilorz V, Steinlechner S. Low reproductive success in Per1 and Per2 mutant mouse females due to accelerated ageing? Reproduction (2008) 135:559–68 https://doi.org/10.1530/REP-07-0434

Pittendrigh CS, Minis DH (1972) Circadian systems: longevity as a function of circadian resonance in Drosophila melanogaster. Proc Natl Acad Sci U S A 69:1537–1539

Prendergast BJ (2008) Behavioral tolerance to endotoxin is enhanced by adaptation to winter photoperiods. Psychoneuroendocrinology 33:540–545

Qian J, Scheer FAJL (2016) Circadian system and glucose metabolism: implications for physiology and disease. Trends Endocrinol Metab 27(5):282–293. https://doi.org/10.1016/j.tem.2016.03.005

Ralph MR, Menaker M (1988) A mutation of the circadian system in golden hamsters. Science 241:1225–1227

Ratajczak CK, Boehle KL, Muglia LJ (2009) Impaired steroidogenesis and implantation failure in Bmal1 mice. Endocrinology 150:1879–1885

Reddy AB, Field MD, Maywood ES, Hastings MH (2002) Differential resynchronisation of circadian clock gene expression within the suprachiasmatic nuclei of mice subjected to experimental jet lag. J Neurosci 22:7326–7330

Reiter RJ, Tan DX, Osuna C, Gitto E (2000) Actions of melatonin in the reduction of oxidative stress. A review. J Biomed Sci 7:444–458

Roenneberg T, Allebrandt KV, Merrow M, Vetter C (2012) Social jetlag and obesity. Curr Biol 22:939–943

Rohling JH, Vanderleest HT, Michel S, Vansteensel MJ, Meijer JH (2011) Phase resetting of the mammalian circadian clock relies on a rapid shift of a small population of pacemaker neurons. PLoS One 6:e25437

Roybal K, Theobold D, Graham A et al (2007) Mania-like behavior induced by disruption of CLOCK. Proc Natl Acad Sci USA 104:6406–6411

Ruby NF, Joshi N, Heller HC (1998) Phase shift magnitude and direction determine whether Siberian hamsters reentrain to the photocycle. J Biol Rhythm 13:506–517

Ruby NF, Hwang CE, Wessells C et al (2008) Hippocampal-dependent learning requires a functional circadian system. Proc Natl Acad Sci U S A 105:15593–15598

Saberi HR, Moravveji AR (2010) Gastrointestinal complaints in shift-working and dayworking nurses in Iran. J Circadian Rhythms 8:9

Sadacca LA, Lamia KA, deLemos AS, Blum B, Weitz CJ (2011) An intrinsic circadian clock of the pancreas is required for normal insulin release and glucose homeostasis in mice. Diabetologia 54:120–124

Sakai T, Tamura T, Kitamoto T, Kidokoro Y (2004) A clock gene, period, plays a key role in long-term memory formation in Drosophila. Proc Natl Acad Sci U S A 101:16058–16063

Salgado-Delgado R, Nadia S, Angeles-Castellanos M, Buijs RM, Escobar C (2010) In a rat model of night work, activity during the normal resting phase produces desynchrony in the hypothalamus. J Biol Rhythm 25:421–431

Sallinen M, Kecklund G (2010) Shift work, sleep, and sleepiness—differences between shift schedules and systems. Scand J Work Environ Health 36:121–133

Santhi N, Horowitz TS, Duffy JF, Czeisler CA (2007) Acute sleep deprivation and circadian misalignment associated with transition onto the first night of work impairs visual selective attention. PLoS One 2:e1233

Scheer FA, Hilton MF, Mantzoros CS, Shea SA (2009) Adverse metabolic and cardiovascular consequences of circadian misalignment. Proc Natl Acad Sci U S A 106:4453–4458

Scheiermann C, Kunisaki Y, Frenette PS (2013) Circadian control of the immune system. Nat Rev Immunol 13:190–198

Sellix MT, Evans JA, Leise TL et al (2012) Aging differentially affects the re-entrainment response of central and peripheral circadian oscillators. J Neurosci 32:16193–16202

Shi SQ, Ansari TS, McGuinness OP, Wasserman DH, Johnson CH (2013) Circadian disruption leads to insulin resistance and obesity. Curr Biol 23:372–381

Shimba S, Ogawa T, Hitosugi S et al (2011) Deficient of a clock gene, brain and muscle Arnt-like protein-1 (BMAL1), induces dyslipidemia and ectopic fat formation. PLoS One 6:e25231

Siepka SM, Yoo SH, Park J, Lee C, Takahashi JS (2007) Genetics and neurobiology of circadian clocks in mammals. Cold Spring Harb Symp Quant Biol 72:251–259

Smarr BL, Morris E, de la Iglesia HO (2012) The dorsomedial suprachiasmatic nucleus times circadian expression of Kiss1 and the luteinizing hormone surge. Endocrinology 153:2839–2850

Spengler ML, Kuropatwinski KK, Comas M et al (2012) Core circadian protein CLOCK is a positive regulator of NF-kappaB-mediated transcription. Proc Natl Acad Sci U S A 109:E2457–E2465

Stephan FK, Zucker I (1972) Circadian rhythms in drinking behavior and locomotor activity of rats are eliminated by hypothalamic lesions. Proc Natl Acad Sci U S A 69:1583–1586

Summa KC, Vitaterna MH, Turek FW (2012) Environmental perturbation of the circadian clock disrupts pregnancy in the mouse. PLoS One 7:e37668

Sun Y, Yang Z, Niu Z et al (2006) MOP3, a component of the molecular clock, regulates the development of B cells. Immunology 119:451–460

Tenkanen L, Sjoblom T, Harma M (1998) Joint effect of shift work and adverse life-style factors on the risk of coronary heart disease. Scand J Work Environ Health 24:351–357

Thomas BB, Oommen MM, Ashadevi JS (2001) Constant light and blinding effects on reproduction of male South Indian gerbils. J Exp Zool 289:59–65

Varcoe TJ, Wight N, Voultsios A, Salkeld MD, Kennaway DJ (2011) Chronic phase shifts of the photoperiod throughout pregnancy programs glucose intolerance and insulin resistance in the rat. PLoS One 6:e18504

Vetter C (2020) Circadian disruption: what do we actually mean? Eur J Neurosci 51(1):531–550. https://doi.org/10.1111/ejn.14255

Wang CY, Wen MS, Wang HW et al (2008) Increased vascular senescence and impaired endothelial progenitor cell function mediated by mutation of circadian gene Per2. Circulation 118:2166–2173

Wang LM, Dragich JM, Kudo T et al (2009) Expression of the circadian clock gene Period2 in the hippocampus: possible implications for synaptic plasticity and learned behaviour. ASN Neuro 1:e00012

Wang XS, Armstrong ME, Cairns BJ, Key TJ, Travis RC (2011) Shift work and chronic disease: the epidemiological evidence. Occup Med 61:78–89

Wise J (2009) Danish night shift workers with breast cancer awarded compensation. BMJ 338:b1152

Wittmann M, Dinich J, Merrow M, Roenneberg T (2006) Social jetlag: misalignment of biological and social time. Chronobiol Int 23(1–2):497–509

Woelfle MA, Ouyang Y, Phanvijhitsiri K, Johnson CH (2004) The adaptive value of circadian clocks: an experimental assessment in cyanobacteria. Curr Biol 14:1481–1486

Wood PA, Yang X, Taber A et al (2008) Period 2 mutation accelerates ApcMin/þ tumorigenesis. Mol Cancer Res 6:1786–1793

Wright KP Jr, Hull JT, Hughes RJ, Ronda JM, Czeisler CA (2006) Sleep and wakefulness out of phase with internal biological time impairs learning in humans. J Cogn Neurosci 18:508–521

Wright KP, Lowry CA, Lebourgeois MK (2012) Circadian and wakefulness-sleep modulation of cognition in humans. Front Mol Neurosci 5:50

Wyse CA, Coogan AN, Selman C, Hazlerigg DG, Speakman JR (2010) Association between mammalian lifespan and circadian free-running period: the circadian resonance hypothesis revisited. Biol Lett 6:696–698

Yang X, Wood PA, Oh EY et al (2009a) Down regulation of circadian clock gene Period 2 accelerates breast cancer growth by altering its daily growth rhythm. Breast Cancer Res Treat 117:423–431

Yang S, Liu A, Weidenhammer A et al (2009b) The role of mPer2 clock gene in glucocorticoid and feeding rhythms. Endocrinology 150:2153–2160

Zee PC, Rosenberg RS, Turek FW (1992) Effects of aging on entrainment and rate of resynchronization of circadian locomotor activity. Am J Phys 263:R1099–R1103

Zeng ZL, Wu MW, Sun J et al (2010) Effects of the biological clock gene Bmal1 on tumour growth and anti-cancer drug activity. J Biochem 148:319–326

Zghoul T, Abarca C, Sanchis-Segura C et al (2007) Ethanol self-administration and reinstatement of ethanol-seeking behavior in Per1Brdm1 mutant mice. Psychopharmacology 190:13–19

Zhao B, Lu J, Yin J et al (2012a) A functional polymorphism in PER3 gene is associated with prognosis in hepatocellular carcinoma. Liver Int 32:1451–1459

Zhao Y, Zhang Y, Zhou M et al (2012b) Loss of mPer2 increases plasma insulin levels by enhanced glucose-stimulated insulin secretion and impaired insulin clearance in mice. FEBS Lett 586:1306–1311

Zhu Y, Stevens RG, Hoffman AE et al (2011) Epigenetic impact of long-term shiftwork: pilot evidence from circadian genes and whole-genome methylation analysis. Chronobiol Int 28:852–861

7 The Human SCN in Health and Age-Related Neurological Disorders

7.1 Introduction

The suprachiasmatic nucleus (SCN) is the primary unit of the mammalian circadian rhythm. The generation and coordination of many physiological, endocrine, and behavioural circadian rhythms are crucial (Morin and Blanchard 2001; Nakagawa and Okumura 2010). The suprachiasmatic nucleus (SCN) in humans, as well as in non-human model systems, acts as the neural basis for circadian rhythms. The disturbance of circadian behaviour, which has been clinically recorded, is linked to the participation of the SCN region (Haugh and Markesbery 1983; Schwartz et al. 1987). The decrease in the ability to keep track of circadian rhythms as one gets older has been linked to the deterioration of SCN neurons during the ageing process (Hofman and Swaab 1994; Mirmiran et al. 1992). The suprachiasmatic nucleus (SCN) is a compact anatomical region in the hypothalamus, positioned bilaterally above the optic nerve and close to the third ventricle. The suprachiasmatic nucleus (SCN) functions as the primary controller of the mammalian brain, coordinating and governing the hormonal and behavioural circadian and circannual rhythms (Swaab 2003). Damage to the suprachiasmatic nucleus (SCN) region, such as from tumours in the suprasellar pituitary or metastases, has been reported to decrease the synthesis of a crucial SCN peptide called arginine vasopressin (AVP) and disturb circadian rhythms (Borgers et al. 2011; Swaab 2003). Hence, it is crucial to employ immunocytochemical labelling methodologies utilising antibodies targeting arginine vasopressin (AVP), vasoactive intestinal polypeptide (VIP), or neurotensin to precisely ascertain the identity of this anatomical feature. Mai and his colleagues employed immunocytochemistry to categorise the human SCN into five primary subgroups (Mai et al. 1991). The region of the suprachiasmatic nucleus (SCN) that receives information from the retinohypothalamic tract (RHT) and is crucial for synchronisation is characterised by the existence of VIP neurons (Moore et al. 2002). AVP is found throughout the rest of the suprachiasmatic nucleus (SCN), whereas neurotensin is evenly distributed throughout the entire SCN. The

S. I. Rizvi, G. Majumdar, *Circadian Clock and Aging*,
https://doi.org/10.1007/978-981-97-6695-6_7

suprachiasmatic nucleus (SCN) consists of fibres that contain somatostatin, thyrotropin-releasing hormone, galanin, preproenkephalin, delta-sleep-inducing peptide, leptin receptor, and hypocretin (Swaab 2003). Furthermore, the human suprachiasmatic nucleus (SCN) has specific sites that bind to vasoactive intestinal peptide (VIP), such as oestrogen receptor (ER) α and β, progesterone receptor (Kruijver and Swaab 2002), and melatonin receptors (MT) (Y.-H. Wu et al. 2013). Considering the existence of VIP in the SCN, it is unsurprising that peptide methionine amide (PHM), a peptide including histidine and methionine, is also present in the human SCN. The PHM and VIP genes are located on adjacent exons of a common prepro-VIP gene. The work conducted by Romijn et al. 1999 employed confocal laser scanning microscopy to identify a small fraction of neurons in the human SCN that showed co-localization with AVP and VIP. Following the application of microwave therapy for antigen retrieval, the detection of AVP and VIP staining exhibits enhanced responsiveness. The quantity of AVP-stained neurons exhibited a 70% rise, whereas the quantity of VIP-stained neurons showed a substantial increase of 280%. The neurons observed before microwave therapy were mainly localised in the central region of the SCN, whereas the neurons that became visible only after microwave treatment were situated in the peripheral area of the SCN. The results suggest that the AVP and VIP neurons in the central part of the SCN have a greater amount of peptide, maybe because they are more active than the peripheral neurons (Hofman and Swaab 1994).

The human suprachiasmatic nucleus (SCN) contains many neurons that express both isoforms of glutamic acid decarboxylase (GAD), specifically GAD65 and GAD67. Gamma-aminobutyric acid (GABA) coexists with one or more peptides in SCN neurons (Swaab 2003). GABA is widely acknowledged as an inhibitory neurotransmitter in the brain. However, its impact may differ based on the circadian cycle. Stimulation of SCN neurons can be achieved by GABA through a mechanism that relies on GABA receptors. The suprachiasmatic nucleus (SCN) of the human foetus and newborn already possesses a moderate concentration of benzodiazepine binding sites. The suprachiasmatic nucleus (SCN) of the human foetus has a densely packed network of catecholaminergic cells, which are visible as early as the third and fourth months of pregnancy.

7.2 The SCN's Role in Neurodegenerative Diseases

Neurodegenerative disease (NDD) is a term used to define a diverse group of illnesses that exhibit different clinical and pathological features. Nevertheless, these illnesses share a fundamental biological characteristic, since they are characterised by the aggregation of misfolded proteins into insoluble compounds, accompanied by a progressive deterioration of nerve cells. An early disruption of the sleep/wake cycle is a typical indication of NDD. Disruptions in sleep patterns and regular physiological indicators frequently happen, indicating that NDDs may impact the circadian timing system. The mammalian circadian timing system consists of a core clock in the suprachiasmatic nucleus (SCN). This clock collects photic information through a specialised channel and also integrates information on other relevant environmental signals, such as the timing of feeding. The suprachiasmatic nucleus (SCN) communicates with other brain regions by many pathways, including arousal centres, the autonomic nervous system, and the hypothalamic-pituitary-adrenal (HPA) axis. The central clock may regulate the body's rhythms by utilising various output pathways, whereas many tissues have their own separate molecular clocks. The result is a thorough temporal framework that affects all biological activities. An essential tenet in the field of circadian biology is that the temporal organisation of our daily rhythms plays a vital role in preserving our well-being (Roenneberg and Merrow 2016). Interrupting these cycles has adverse effects that impact multiple elements of physiology. This chapter delineates several adverse outcomes that result from a dysfunctional temporal framework. Fig. 7.1 illustrates the association between various neuro-disorders and diseases and the normal operation of the circadian clock.

7.2.1 Alzheimer's Disease (AD)

The human ageing process is linked to changes in circadian rhythms and a higher prevalence of sleep disorders (Van Someren et al. 2002). The modifications in the neuronal structure of the SCN, the diminished amount of light input to the internal

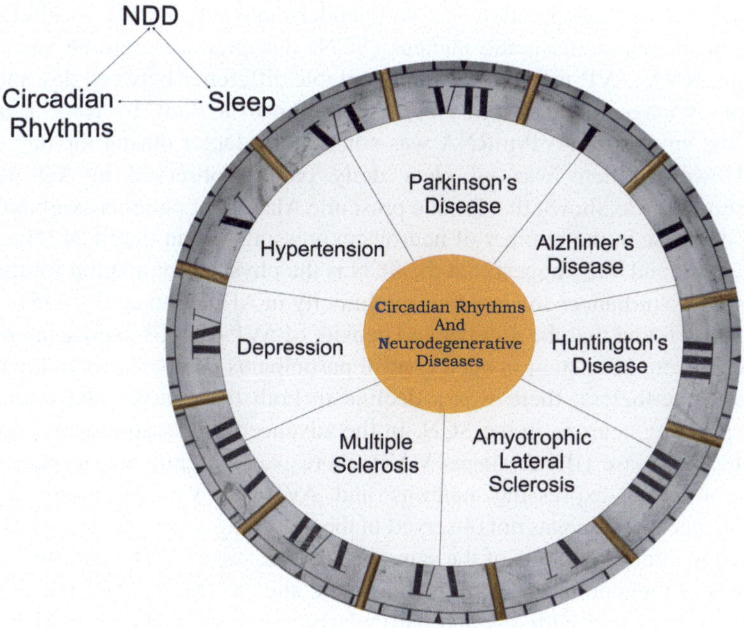

Fig. 7.1 Depiction of multiple neurological disorders/conditions linked to the proper functioning of clock

clock, and the notable decline in peptide synthesis in the SCN are observed in patients with Alzheimer's Disease (AD) (R.-Y. Liu et al. 2000; Swaab 2004). Moreover, the presence of pretangles (Van de Nes et al. 1998) and tangles (Stopa et al. 1999) in the SCN of persons with Alzheimer's disease suggests that the disease affects this structure as it progresses. Nevertheless, the occurrence of scattered amyloid plaques in this nucleus is infrequent. Research has shown that the suprachiasmatic nucleus (SCN) in Alzheimer's disease (AD) patients, who often suffer from sleep difficulties and abnormalities in their biological clock, is severely affected (Stopa et al. 1999; Van de Nes et al. 1998). This condition is marked by a notable reduction in neurons that produce arginine vasopressin (AVP) and neurotensin and a corresponding increase in astrocytes that are stained with glial fibrillary acidic protein (GFAP). Harper and colleagues discovered that a clear decrease in SCN neurotensin neurons, observed in postmortem material, was associated with a decrease in activity and temperature range, without an increase in activity disruption, when compared to clinical data collected before death. Moreover, the reduction of SCN AVP-expressing neurons was associated with increased activity fragmentation but not with a decrease in amplitude. The study conducted by Harper and colleagues provides valuable insights into the role of certain groupings of neurons in supporting the function of the SCN and the potential benefits of AD (Harper et al. 2008).

The immunocytochemical data demonstrating decreased functionality of the suprachiasmatic nucleus in Alzheimer's disease have been confirmed using in situ hybridization. The level of AVP-mRNA in the SCN of AD patients was significantly decreased by thrice compared to age- and gender-matched controls. In addition, the neurons in the suprachiasmatic nucleus (SCN) that produce arginine vasopressin messenger RNA (AVP-mRNA) showed a notable difference between day and night in persons younger than 80 years old. According to a study by R.-Y. Liu et al. (2000), the amount of AVP-mRNA was consistently larger during the day than at night. However, there was no clear daily pattern observed in AD patients. Furthermore, it was shown that female presenile Alzheimer patients exhibited a significant decrease in the number of neurons expressing VIP in the SCN. The results from the AVP and VIP suggest that the SCN is the physical foundation for the commonly seen disturbances in circadian rhythmicity in AD (Zhou et al. 1995).

The study found that the number and density of AVP- or VIP-expressing neurons in the SCN were consistent in aged control participants compared to young control subjects. Nevertheless, there was a decline in both the number and intensity of MT1-expressing neurons in the SCN. In the advanced neuropathological stages of Alzheimer's disease (Braak stages V-VI), there was a significant decrease in the presence of MT1-expressing neurons and AVP- or VIP-expressing neurons. However, this decrease was not observed in the early stages (Braak stages I-II) when compared to control patients of the same age (Braak stage 0). The data suggest that the impact of melatonin on the suprachiasmatic nucleus (SCN), facilitated by MT1 receptors, is impaired with age and particularly in advanced stages of Alzheimer's disease (AD). This disturbance may lead to the emergence of clinical circadian irregularities and impact the efficacy of therapeutic melatonin treatment in these conditions (Wu et al. 2006).

Alzheimer's disease (specifically, Braak stages II-III) already disrupts the production of melatonin in the pineal gland at an early stage. The disruption is thought to be due to a malfunction in the suprachiasmatic nucleus (SCN). The pineal gland does not exhibit daily fluctuations in the expression of clock genes hBmal1, hCry1, hPer1, and the β-adrenergic receptor. Moreover, the levels of AVP-mRNA decrease in the suprachiasmatic nucleus (SCN) of both preclinical (Braak stages I-II) and clinical (Braak stages V-VI) individuals with Alzheimer's disease. According to Y. Wu et al. (2006), the research indicates that the functioning of the SCN and pineal gland is reduced during the initial phase of Alzheimer's disease (AD) progression. Research conducted in the Netherlands with a sample size of 189 individuals, 87% of whom were diagnosed with dementia, demonstrated that the administration of both light and melatonin treatment resulted in a substantial enhancement in sleep efficiency and a decrease in nocturnal restlessness (Riemersma-van Der Lek et al. 2008). Unlike the utilisation of hypnotics, these therapies for sleep disorders result in enhanced sleep quality without any adverse effects. Moreover, they also improve mood, enhance performance, boost daily energy levels, and promote general quality of life. The existing research indicates that increased brain stimulation has the ability to improve or restore reduced neuronal activity.

In summary, both the SCN and the pineal gland experience a decrease in their functionality as individuals age and in the early stages of Alzheimer's disease, prior to the manifestation of symptoms. Nevertheless, these alterations in the circadian system become increasingly apparent in the advanced stages of Alzheimer's disease. Increased ambient light can efficiently activate the suprachiasmatic nucleus (SCN) in persons with Alzheimer's disease (AD).

7.2.2 Huntington's Disease (HD)

Huntington's disease (HD) is a lethal hereditary condition distinguished by a repeated sequence of cytosine-adenine-guanine (CAG) in exon 1 of the Huntingtin gene (HTT) (The Huntington's Disease Collaborative Research Group, 1993). The extended sequence contains a mutant version of the huntingtin protein (mHTT), which is associated with widespread molecular and cellular abnormalities, significant malfunction and neuronal loss (Kim et al. 2016), and the development of both motor and non-motor symptoms. The symptoms encompass a gradual decline in motor coordination, cognitive functions, and mood (Tabrizi et al. 2020). Data indicates that individuals may encounter disruptions in their sleep patterns, abnormalities in their autonomic nervous system, and irregularities in their metabolic processes. Sleep disruptions encompass challenges in achieving restful sleep, as evidenced by many studies (Arnulf et al. 2008; Lazar et al. 2015; Piano et al. 2015). Autonomic abnormalities pertain to dysfunctions in biological processes regulated by the autonomic nerve system, including excessive perspiration, urinary problems, swallowing difficulties, sexual dysfunction, and alterations in heart rate variability

(Andrich et al. 2002; Kobal et al. 2012). Some investigations have detected metabolic anomalies, such as aberrant metabolic processes (for further information, refer to Browne 2008). In addition, studies have shown that cognitive and sleep disorders might appear many years before the first signs of motor dysfunction (Lazar et al. 2015; Tabrizi et al. 2020). The intricate development and advancement of these clinical characteristics in HD entail alterations in neuroendocrine signalling, particularly the release of cortisol and melatonin, as well as disturbances in circadian rhythmicity. The modifications indicate that the operation of the hypothalamic-pituitary-adrenal (HPA) axis and the suprachiasmatic nucleus (SCN) is compromised in HD. For further information, please consult table 1. The idea is backed by neuropathological changes, including a decrease in brain volume, the loss of neurons that release orexin, and reduced levels of vasoactive intestinal peptide (VIP) and arginine vasopressin (AVP) proteins in the hypothalamus (Politis et al. 2008; Soneson et al. 2010).

The activity of the suprachiasmatic nucleus (SCN) is coordinated with the diurnal cycle of light and darkness in the surrounding environment. The synchronisation occurs via the retinal-hypothalamic tract and indirectly through the retinogeniculate circuits. The suprachiasmatic nucleus (SCN) subsequently transmits this information to further hypothalamus nuclei, the reticular formation, and the pineal gland. This coordination guarantees that the daily activities of these brain regions are synchronised (Brzezinski et al. 2021). Melatonin controls the body's internal clock in response to the daily cycle and initiates the process of maintaining body temperature, resulting in a fall in core body temperature and the beginning of feeling sleepy (Brzezinski et al. 2021; Kräuchi et al. 2002). The variation in core body temperature across a 24-h cycle is associated with the internal structure of sleep, namely the circadian rhythm of rapid eye movement (REM) sleep (Dijk and Czeisler 1995). The disturbance of circadian rhythmicity in HD suggests a possible two-way connection with the advancement of the neurodegenerative disease (Videnovic et al. 2014). The basis for this concept is obtained from transgenic animal models of Huntington's disease (HD), namely the R6/2, R6/1, and BACHD mice. These studies have shown that circadian disturbance occurs prior to the appearance of illness symptoms (Kudo et al. 2011; Maywood et al. 2010). Furthermore, individuals with Huntington's disease experience disturbances in their circadian rhythm, which are marked by irregularities in their sleep and activity patterns, abnormal day and night ratios, shifts in sleep and wakefulness timing, and changes in the levels of melatonin and cortisol (Videnovic et al. 2014). The neurological mechanisms responsible for these alterations have not yet been fully understood. Various investigations have shown that alterations in the SCN are highly significant in the disruption of circadian rhythms in HD. In the R6/2 mouse model of Huntington's disease (HD), the normal transcription of core clock genes in the suprachiasmatic nucleus (SCN) and other brain areas is disrupted in living beings. Nevertheless, when scrutinised in controlled laboratory conditions, this disruption is counteracted, suggesting that circadian impairments are a result of alterations in the internal wiring of the SCN (Maywood et al. 2010). Postmortem histopathological analysis reveals a decrease in protein levels of VIP and AVP within the suprachiasmatic nucleus (SCN) of patients with Huntington's disease (HD), as demonstrated in the study by van Wamelen et al. (2021). This reduction in VIP levels in transgenic animal models of HD is associated with disruptions in the circadian rhythm, as indicated by the research conducted by Fahrenkrug et al. (2007). The disruption of circadian rhythmicity can affect several molecular, cellular, and physiological processes, such as sleep (Archer et al. 2014; Lazar et al. 2015). Disruptions in sleep can negatively impact human physiology in several ways. These include impaired brain function and reduced brain size (Joo et al. 2014), disturbances in metabolism (Schmid et al. 2015), dysregulation of the immune system (Irwin 2002), compromised cardiovascular health (Zhong et al. 2005), impaired cognitive abilities (Lo et al. 2014), and mood disturbances (Dinges et al. 1997). There is a high likelihood that sleep and circadian abnormalities are interrelated in Huntington's disease (HD). The presence of mutant huntingtin (mHTT) results in a progressive decline in neuronal function and the subsequent demise of cells in both the striatal and extra-striatal regions, including the hypothalamus nuclei (Tabrizi et al. 2020). Studies conducted on mice have shown significant hypothalamus degeneration and anomalies in the pituitary and adrenal glands (Björkqvist et al. 2008; Du et al. 2017). Research utilising postmortem and structural imaging methods has revealed that patients with HD mutations undergo a reduction in hypothalamic volume. More precisely, there is a notable decrease in size in the hypothalamic nuclei, a reduction in the number of neurons (especially in the nucleus tuberalis lateralis and lateral hypothalamus), and an increase in the activity of microglial cells. Autopsy investigations have also observed the reduction of orexin-releasing neurons, which play a crucial role in providing nerve fibres to the suprachiasmatic nucleus (SCN), in the hypothalamus of persons with Huntington's disease (HD) (Aziz, Jurgens, et al. 2009a). The reduction of these clusters of neurons is thought to contribute to disturbances in the circadian rhythm, dysfunction of the HPA-axis, and subsequent alterations in the secretion of cortisol (Aziz, Jurgens, et al. 2009a). Moreover, the disruption of the HPA-axis has been proposed as a potential underlying factor for the simultaneous occurrence of depression in neurodegenerative illnesses, such as HD.

Remarkably, a study conducted by Gabery et al. (2015) revealed that individuals with preHD did not exhibit any reduction in hypothalamic volume after a 12-month follow-up MRI. The conflicting findings underscore the importance of precisely delineating the changes in volume in the hypothalamus and other brain regions in people with preHD. It is imperative to examine whether these alterations are accountable for or contribute to circadian rhythm disturbances and clinical manifestations reported in HD. Research has indicated that people with Huntington's disease (HD) have significantly lower amounts of melatonin on average and undergo changes in the timing of their melatonin cycle (acrophase). Individuals with preHD also exhibit a propensity for reduced melatonin levels, but this inclination is not as prominent (Kalliolia et al. 2014). Individuals with HD mutations have been discovered to undergo a change in the timing of melatonin synthesis, potentially explaining the sleep abnormalities that have been seen. Studies have shown that individuals with HD have a delay in the initial increase of melatonin in the morning. The delay in the generation of melatonin helps to clarify the observed delay in the sleep-wake time of these patients (Aziz et al. 2009b). The precise aetiology of the decrease or delayed production of melatonin observed in HD is not thoroughly comprehended. However, it may be associated with the progressive dysfunction of neurons in the SCN (Kalliolia et al. 2014). In addition, individuals with HD have experienced sleep disruptions, as indicated by various studies that have reported symptoms such as insomnia, reduced REM sleep, decreased slow wave sleep and sleep efficiency, advanced sleep phase, frequent awakenings during the night, and increased periodic leg movements (PLMs). For more information, refer to the studies conducted by Neutel et al. (2015) and Piano et al. (2015). Furthermore, numerous other research (refer to Lazar et al. 2015 for more information) have shown that individuals with preHD experience sleep disruption, characterised by a fragmented sleep pattern and a decrease in theta power during REM sleep. The presence of these attributes was associated with the intensity of the illness. The presence of sleep disorders aligns with the emergence of cognitive deficits, indicating a potential association between these aspects of illness.

Sleep abnormalities in persons with HD are a notable issue due to their potential to hinder cognitive abilities associated with the hippocampus and PFC, even prior to the onset of the disease (Tabrizi et al. 2020). Disruptions in sleep patterns and increased levels of glucocorticoids can negatively impact these cognitive abilities (Belanoff et al. 2001; Oei and Everaerd 2006). In summary, disruptions in the body's internal clock caused by changes in the molecular clock and problems in the hypothalamus and SCN can lead to changes in the release of melatonin and increased levels of cortisol. This can cause sleep problems, which can worsen the progression of neurodegenerative diseases like Huntington's disease. These sleep problems can also contribute to cognitive and motor deficits as well as mood disturbances in individuals with Huntington's disease (Table 7.1).

7.2.3 Parkinson's Disease (HD)

Parkinson's disease (PD) is distinguished by a specific combination of movement issues, such as reduced movement (bradykinesia), increased tremors at rest, and rigidity with diminished reflexes for maintaining posture (Williams-Gray et al. 2007). However, there are also extensively documented non-motor symptoms that contribute to the slow decline of cognitive function, ultimately leading to dementia (Pontone et al. 2013; Williams-Gray et al. 2007). The diseases include metabolic abnormalities, alterations in olfactory perception, dysfunction of the cardiovascular system, issues in the gastrointestinal tract, and, specifically addressed in this analysis, sleep disorders and disturbances (Menza et al. 2010; Ziemssen and Reichmann 2010). An important finding is that the manifestation occurs several years before the onset of motor symptoms (Willison et al. 2013).

Parkinson's disease (PD) is thought to arise from the degeneration of dopaminergic neurons in the substantia nigra pars compacta (SNpc), which is attributed to Lewy body pathology. Therefore, neurons experience an abnormal accumulation of α-synuclein. Observable indications become evident when there is sufficient

Table 7.1 Types of HD pathologies related to circadian clock

Pathology type	Evidence and signs
Disturbances in circadian rhythms	***Evidence*** 1. Disruption in the pattern of rest and activity 2. Disruption of the normal regulation of gene expression that follows a circadian rhythm, specifically in the suprachiasmatic nucleus (SCN) 3. Decreased circadian regularity of inherent electrical activity in SCN neurons ***Signs of HD*** 1. Disturbances in rest-activity profiles 2. Abnormal day-night ratios 3. Altered sleep-wake timing
Disturbances in melatonin rhythms	***Evidence*** 1. Temporal distribution of melatonin increase ***Signs of HD*** 1. Reduced average and peak melatonin levels 2. Temporal distribution of the increase in melatonin levels 3. Phase delayed increase
Disturbances in sleep rhythms	***Evidence*** 1. Disrupted sleep pattern 2. Reduced theta power seen during REM sleep ***Signs of HD*** 1. Sleeplessness 2. Reduced REM sleep 3. Reduced slow wave sleep (SWS) 4. Reduced sleep efficacy 5. Sleep phase advanced beyond normal 6. Regular occurrences of waking up during the night

deterioration in the nigral region, leading to the manifestation of motor symptoms (Boeve 2013). However, it is becoming increasingly evident that this is a complex illness that affects numerous brain systems as it advances (Willison et al. 2013). This is apparent in the reduced functioning of the extrapyramidal system, which consequently hinders the processes responsible for maintaining alertness and finally results in the emergence of sleep disorders. Individuals with Parkinson's disease (PD) do not experience any damage to the suprachiasmatic nucleus (SCN), which functions by forming connections with the hypothalamus and the central autonomic network linked to the brainstem. Dysfunctional effects arise from the presence of defective circuits (Willison et al. 2013), leading to subsequent alterations in circadian rhythmicity. Consequently, there is a decline in the quality of sleep at night, less alertness throughout the day, and poorer cognitive function (Buysse 2005). Individuals with Parkinson's disease (PD) showed a lack of time-dependent variations in Bmal1 expression, in contrast to the control group. The study found that patients had abnormal sleep patterns, including prolonged time to fall asleep, reduced sleep quality, and decreased REM sleep. This pertains to certain hormone patterns in the bloodstream that are associated with fluctuations in the expression of clock genes in peripheral tissues (Breen et al. 2014). According to the existing knowledge, sleep-wake disturbances in the early stages of Parkinson's disease suggest a malfunction in the brain circuitry responsible for regulating circadian rhythms. Studies have demonstrated that Parkinson's disease (PD) impacts the hypothalamus, resulting in reduced activity of the suprachiasmatic nucleus (SCN). This decline in activity leads to reduced melatonin production and changes in sleep-wake rhythms. Moreover, studies have shown that mice with higher levels of α-synuclein experience a reduction in the rate at which the suprachiasmatic nucleus (SCN) fires, leading to a disruption in the transmission of neuronal and hormonal signals from the central clock (Kudo et al. 2011). A possible explanation for this occurrence is the removal of regulations on the expression of the Bmal1 gene. Dopamine regulates the functioning of the Bmal1/Clock heterodimer (Dibner et al. 2010). Therefore, a dopamine deficiency can directly influence this essential component of the molecular clock. The dopamine metabolism in the striatum is seemingly controlled by clock proteins, including Per2 (Hampp et al. 2008). Dopamine receptor activation also affects the timing of clock gene transcripts of Per1 and Per2 in the striatum (Hood and Amir 2017). Dopamine regulates the consistent pattern of melanopsin expression in retinal ganglion cells, which subsequently impacts the synchronisation of the circadian cycle with light (Sakamoto et al. 2005). Considering the important function of dopamine in controlling the sleep-wake cycle, it is highly probable that the complex connections between different brain structures involved in sleep regulation, like the suprachiasmatic nucleus (SCN), can indefinitely lead to different types of sleep disorders. Alternatively, the disruption of clock genes, or perhaps the hypothalamus, may be caused by damage and deterioration of the SCN itself. One may question whether circadian disruption in Parkinson's disease (PD) can be linked to the underlying mechanisms of other circadian sleep disorders, such as advanced sleep phase syndrome (ASPS), delayed sleep phase syndrome (DSPS), non-24-h sleep-wake syndrome, and seasonal affective disorder. This is because certain symptoms of these disorders overlap with sleep abnormalities associated with Parkinson's disease. Further inquiry is required to determine the exact pathways responsible for the development of these disorders (Cermakian and Boivin 2003). In addition, Anti-parkinsonian medications have the ability to modify circadian rhythms and sleep-wake regulating systems. This can occur either by directly affecting the sleep-wake regulating systems and circadian rhythm generators or indirectly targeting PD symptoms to alleviate them during sleep (Garcia-Borreguero et al. 2004). The study found that the pineal gland undergoes compensatory up-regulation of monoaminergic transmitter networks in regions outside of the basal ganglia. This particularly happens in its absorption of L-DOPA, which is the precursor of DA. The main factor influencing this behaviour is the extended use of L-DOPA, dopamine agonists, or MAO-B inhibitors (Ghaemi et al. 2001). These drugs alter the usual physiological mechanisms related to the breakdown of DOPA in the pineal gland, resulting in an enhanced uptake of L-DOPA. This phenomenon is caused by an increase in the number of binding sites of striatal dopamine uptake carriers, which are not only present in the dopaminergic system but also in the adrenergic and serotonergic systems. Therefore, the disruption of pineal gland function is not exclusively attributed to metabolic irregularities in the dopaminergic systems but also to the adrenergic and serotonergic systems.

Consequently, this leads to a disturbance in the circadian rhythm as a result of the pineal gland's role in the synthesis of melatonin, which regulates the sleep-wake cycle. A study found more data showing that Parkinson's disease (PD) patients who were using L-DOPA medication had an earlier evening melatonin peak compared to newly diagnosed patients and healthy individuals (Nagatsu and Sawada 2009). Individuals diagnosed with Parkinson's disease (PD) experience reduced sleep duration and poorer sleep quality, which is marked by more instances of waking up during the night and heightened alertness after falling asleep (Wetter et al. 2000). Sleep fragmentation is a common sleep disturbance characterised by frequent awakenings and remaining awake for 30–40% of the total sleep time. Several studies have shown that there is a decrease in sleep spindles and slow-wave sleep (SWS) (Garcia-Borreguero et al. 2004), as well as a shorter time to enter rapid eye movement sleep (REMS) (Poewe and Högl 2000). However, increased arousals are associated with excessive daytime sleepiness (EDS). These symptoms suggest changes in the timing of sleep caused by disturbances in the body's natural circadian rhythm (for more information, see Schulte and Winkelmann 2011). The disruption of the circadian rhythm is a significant factor that contributes to both insomnia and hypersomnia in patients with Parkinson's disease (Willison et al. 2013). Patients with Parkinson's disease (PD) also experience rapid eye movement sleep (REMS) behaviour disorder (RBD) as a result of structural abnormalities in the brainstem caused by a decrease in dopamine levels and impairments in neural circuits. The brainstem is the primary region responsible for regulating REMS. These studies collectively offer data supporting the association between circadian dysfunctions and Parkinson's disease (PD). However, the underlying reason for the circadian symptoms in individuals with Parkinson's disease remains incompletely understood. The challenge originates from various factors. Firstly, the disorder itself is characterised by a high degree of diversity, with a broad spectrum of symptoms that demonstrate significant variation (Thenganatt and Jankovic 2016). In addition, the treatment for Parkinson's disease (PD) can have a separate effect on circadian parameters, such as the timing of melatonin cycling. This is especially true for patients taking levodopa medication (Bolitho et al. 2014). In addition, sleep problems, which get more severe as the disease advances, have a harmful impact on the circadian system of individuals with Parkinson's disease, leading to a self-perpetuating loop (Fifel and Videnovic 2019). Both circadian and sleep characteristics have been studied in different mammalian models of Parkinson's disease, which were developed by genetic changes or toxin-induced treatments. In this context, our main objective is to identify flaws predominantly associated with the circadian clock. These defects are listed in Table 7.2.

7.2.4 Amyotrophic Lateral Sclerosis (ALS)

Amyotrophic lateral sclerosis (ALS) is a complex degenerative condition marked by the degeneration of upper and lower motor neurons, leading to a variety of motor and non-motor symptoms (Hardiman et al. 2017). The cause of ALS is mostly unknown; nevertheless, it is proven that it is associated with genetic changes. The

Table 7.2 Disruptions in the circadian rhythm in animal models of Parkinson's disease (PD)

Species	PD model	Effects on clock
Mice	ASO	The circadian rhythms are disrupted and the firing rate of neurons in the SCN is decreased during the daytime
	Mito-Park	Age-related decrease in rhythmicity and disrupted circadian activity patterns in the presence of continual high levels of light
	MPTP-injection	Reduced amplitude in locomotor rhythm; altered clock Gene expression in the SCN
Rat	6-OHDA-injection	1. Modified gene expression of clock genes in the striatum2. In a condition of constant darkness, there is an irregular pattern of wheel-running behaviour and a decrease in the increase of PER2 expression in the dorsal striatum3. The locomotor activity period is decreased in LD conditions, and the activity rhythm has a longer periodicity under constant dim light4. Reduced magnitude of the heart rate pattern
Dog	MPTP-injection	Changes in urine volume and vasopressin release over a 24-hour period
Non-human primates	MPTP-injection	Absence of light/dark cues leads to the loss of circadian locomotor activity

expansion of hexanucleotide repeats in C9orf72, a gene that encodes the C9orf72 guanine nucleotide exchange factor, is recognised as a key contributor to familial ALS (Gossye et al. 2020). This gene functions as a guanine nucleotide exchange factor, which is responsible for the regulation of specific Rab GTPases. Moreover, a connection exists between amyotrophic lateral sclerosis (ALS) and genetic modifications in the superoxide dismutase 1 (SOD1) gene (Berdyński et al. 2022). While motor symptoms are the primary features of the disease, sleep alterations are also documented and are linked to the advancement of symptoms (Boentert 2020). In amyotrophic lateral sclerosis (ALS), like other neurodegenerative illnesses, there is a noticeable variation in significant physiological markers, such as body temperature (Braun et al. 2019). The ALS transgenic mice exhibited reduced body temperature variability during both the light and dark phases, a trait that was not observed in the control animals. It is noteworthy that control mice display elevated body temperature during the dark phase, which corresponds to their activity period. The rise in temperature reduces the energy consumption resulting from cold weather. The disruption of the circadian rhythm seems to play a part in the progression of ALS (Huang et al. 2018). This study discovered that in mice afflicted with ALS, the disturbance of circadian rhythm resulted in an earlier commencement and more rapid advancement of the disease. This was demonstrated by the presence of more severe functional limitations, reduction in body weight, higher loss of motor neurons, heightened activation of glial cells, and inflammation in the spinal cord caused by nuclear factor κB. Additionally, an increase in the occurrence of intestinal cyanobacteria was seen in ALS transgenic mice. Prior to the onset of cognitive deterioration in ALS, it is hypothesised that disturbances in the sleep-wake cycle and circadian rhythm take place (Zhang et al. 2020). According to this study, mice with a genetic mutation that leads to ALS exhibited disruptions in their circadian rhythm and sleep habits prior to the onset of any cognitive or motor abnormalities. This finding suggests that the study of circadian behaviour could be a useful physiological method for identifying and treating diseases, such as ALS, at an early stage. The expression of clock genes in ALS is modified. For instance, in transgenic mice with ALS, the gene Fus, which regulates circadian rhythm and typically exhibits diurnal oscillating expression, demonstrated decreased oscillation. The aberrant expression of Fus has a detrimental effect on the well-being of neurons and also has an influence on the expression of other clock genes, such as hypocretin receptor 2 (Hcrtr2) and many ankyrin repeat domains 3 (Shank3), as determined by the zeitgeber period (Zhang et al. 2020).

In addition to the altered expression of clock genes, the expression of clock-controlled genes is also modified in ALS (Killoy et al. 2021). The expression of Nr1d1, a pivotal regulator of the molecular clock that mediates communication between circadian rhythm, metabolism, and inflammation, was reduced in transgenic mice with ALS (Killoy et al. 2021). Observations have shown that the decrease in Nr1d1 expression in astrocytes leads to a detrimental phenotype in motor neurons, potentially establishing a connection to ALS. Furthermore, the activity of genes that produce sirtuins (Sirt1, Sirt3, and Sirt6), metabolic enzymes (Pfkfb3, Cpt1, and Nampt), and redox regulators (Nrf2, G6pd, Pgd)—all of which are controlled by clock genes—was similarly reduced. These findings indicate that, in addition to Nr1d1, other clock genes are also reduced in ALS (Killoy et al. 2021). In addition, the expression of PER2 is reduced in fibroblasts and skeletal muscle cells of persons diagnosed with ALS (Mamoor 2022). This finding is fascinating since this gene plays a crucial role in the mechanism by which the circadian clock in vertebrates reacts to light and can lead to diverse alterations in the molecular circadian clock, including disruptions in the circadian rhythm. ALS is a neurodegenerative disease that affects motor neurons and is associated with alterations in the expression of clock genes and clock-controlled genes. Transgenic mice with ALS displayed reduced oscillation and expression of the circadian rhythm regulator genes Fus and Nr1d1, which play a vital role in controlling the molecular clock. Furthermore, the transgenic mice with ALS demonstrated a reduction in the expression of clock-controlled genes, such as Sirt1, Sirt3, Sirt6, Pfkfb3, Cpt1, Nampt, Nrf2, G6pd, and Pgd (Fig. 7.2). Moreover, the level of PER2 expression is reduced in fibroblasts and skeletal muscle cells of persons diagnosed with ALS. These data

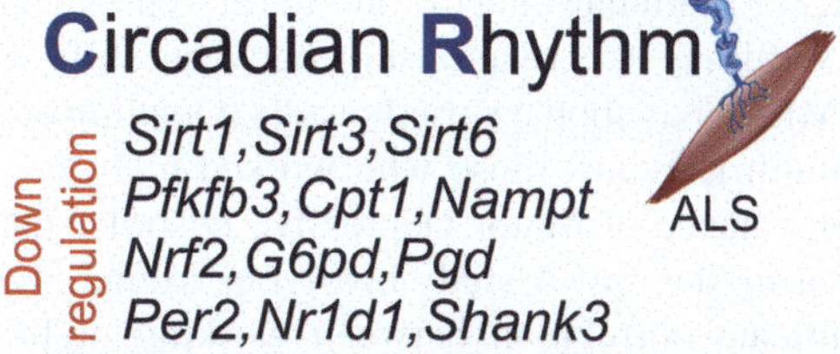

Fig. 7.2 Circadian clock and Amyotrophic Lateral Sclerosis (ALS): genes involved

suggest that the disturbance of the circadian rhythm has a role in the progression of ALS. Further investigation is necessary to elucidate the mechanisms underlying circadian rhythm disruption in ALS and explore potential therapeutic options.

7.3 Depression

There is a reciprocal relationship between mood disorders and circadian rhythms. Mood disorders often arise from disruptions in circadian clock-regulated systems, such as sleep patterns and cortisol secretion. Conversely, the disturbance of natural sleep-wake cycles due to circumstances such as jet lag, working night shifts, or being exposed to artificial light during nighttime can initiate or exacerbate emotional problems in vulnerable individuals. Substantial data exists that demonstrates strong links between circadian rhythms and mental health. Only recently has research begun to reveal the direct interrelationships between the circadian system and the regulation of mood. Depression, also known as Major depressive disorder (MDD), is characterised by alterations in mood, typically characterised by increased sadness and/or irritability, accompanied by at least one of the following psycho-physiological symptoms: changes in sleep patterns, sexual desire, or appetite, inability to experience pleasure, slowed speech or movements, crying, and suicidal thoughts (Belmaker and Agam 2008). To diagnose Major Depressive Disorder (MDD), these symptoms must be present for at least 2 weeks and have a considerable impact on regular daily activities. Major Depressive Disorder (MDD) exerts a substantial influence on a vast global population. A study undertaken by the Global Burden of Disease Consortium examined the frequency and extent of 328 diseases across 195 nations from 1990 to 2016. MDD was identified as a leading contributor to the disease burden in nearly all countries, with only four exceptions. The global prevalence of Major Depressive Disorder (MDD) is increasing, with a significant 18% rise in the number of individuals diagnosed with depression between 2005 and 2015 (No. WHO/MSD/MER/2017.2; World Health Organisation, 2017). The occurrence of Major Depressive Disorder (MDD) is closely linked to the degree of modernization in society (Hidaka 2012). This phenomenon can be ascribed to the increased disruption of circadian rhythms, which can be caused by causes such as exposure to artificial light at night, shift work, and jet lag. Additionally, it may also be influenced by the interaction between circadian disruption and other environmental elements prevalent in modernised countries. There is a scarcity of human studies that have particularly examined the correlation between disruptions in the circadian rhythm and Major Depressive Disorder (MDD). This is due to the tendency of human data to combine all types of depression into one aggregate. Research examining the relationship between shift work and Major Depressive Disorder (MDD) has shown varied results (Oenning et al. 2018). Among approximately 4000 South Koreans, those who worked night or shift shifts had a significantly greater prevalence of Major Depressive Disorder (MDD) compared to those who worked during the day. A study involving around 36,000 Brazilian workers revealed a significant correlation between working night shifts and the occurrence of Major

Depressive Disorder (MDD), particularly among females. A meta-analysis of 11 studies found that night shift workers have a 40% higher risk of getting depression compared to those who work during the day (Lee et al. 2017).

There is a scarcity of studies about the influence of jet lag or social jet lag on Major Depressive Disorder (MDD). A specific study examined the extent of social jet lag, defined as the absolute disparity in midsleep duration between days off and work days, in individuals diagnosed with Major Depressive Disorder (MDD) and in individuals without any mental health conditions. Furthermore, persons diagnosed with Major Depressive disease (MDD) did not show any differences in social jet lag when compared to individuals without the disease (Knapen et al. 2018). However, like shift work, the process of categorising depression by combining various types of distress creates an association between jet lag and depression, including social jet lag (McNeely et al. 2018). Human clinical research provides empirical evidence that strongly supports a significant association between Major Depressive Disorder (MDD) and circadian rhythms. Depression symptoms exhibit diurnal variations, with patients feeling either worsened symptoms in the morning or worsened symptoms in the evening (Rusting and Larsen 1998). People diagnosed with Major Depressive Disorder (MDD) sometimes experience more severe symptoms in the early hours, suggesting a stronger expression of depression (Rusting and Larsen 1998). Major Depressive Disorder (MDD) is primarily characterised by disruptions in biological cycles. These disruptions encompass alterations in sleep patterns (such as shortened time to enter rapid eye movement sleep, increased rapid eye movement sleep, and decreased slow wave sleep), disturbances in social rhythms, changes in hormone rhythms (such as decreased melatonin and cortisol levels), and disruptions in body temperature rhythms (such as decreased intensity and increased nighttime body temperature). These disturbances are frequently reported in persons with Major Depressive Disorder (MDD) (Vadnie and McClung 2017). Empirical evidence suggests that the severity of Major Depressive Disorder (MDD) is directly correlated with the degree of disturbance in circadian rhythms (Emens et al. 2009). In addition, examination of the daily patterns of gene activity in the brains of deceased individuals with Major Depressive Disorder (MDD) shows reduced strength, altered peak times, and changed coordination between genes, particularly those involved in regulating the body's internal clock. The human suprachiasmatic nucleus (SCN) displays both diurnal and annual fluctuations in neuronal activity, which is quite remarkable. The aforementioned variances are thought to be linked to alterations in mood and sleep disturbances in individuals suffering from depression (van Londen et al. 2001). Depression symptoms exhibit fluctuations over the course of the day, and the body's stress response is likewise influenced by the time of day. The function of the suprachiasmatic nucleus (SCN) is directly influenced by light, and light therapy has been reported to affect depression. Seasonal affective disorder (SAD) is more prevalent in the northern regions of the United States in comparison to the southern states (Miller et al. 2005). A robust association exists between light, the suprachiasmatic nucleus (SCN), and mood. Considering the connection between genetic variations in clock genes and GCK3, which is affected by lithium, and the resulting impact on disrupted circadian rhythm and increased susceptibility to mood disorders such as seasonal affective disorder (SAD) and bipolar disorder (BD) (Waddington Lamont et al. 2007), it is plausible that the suprachiasmatic nucleus (SCN) may play a role in the onset of depression, particularly in specific groups of patients. The diminished activity of the SCN in depression is thought to be caused, at least partially, by the heightened levels of cortisol in the circulatory system of patients with depression. The reason for this is that corticosteroids inhibit the synthesis of AVP mRNA in the SCN (A. C. Liu et al. 2007). An extensive examination of the complete set of genes expressed in the human brain indicated that patients suffering from serious depression displayed reduced diurnal fluctuations. This phenomenon was ascribed to a change in the time of the highest level of activity and a potential disturbance in the synchronisation of some circadian genes.

In individuals with depression, it was revealed that the quantity of MT1-immunoreactive (ir) cells and AVP- and/or VIP-ir cells in the central SCN, which is the suprachiasmatic nucleus, was higher. This indicates an increased presence of melatonin and certain neuropeptides in the brains of depressed individuals. Nevertheless, there was no alteration in the quantity of MT2-ir cells. Moreover, there was an inverse relationship between the number of MT1-ir cells and the age of onset of depression, whereas there was a direct relationship with the duration of the condition. Elevated levels of MT1 receptors were observed specifically in the suprachiasmatic nucleus (SCN) of persons with depression. Furthermore, these levels may continue to grow as the illness progresses. These alterations could be implicated in disruptions of the circadian rhythm and might have a role in the effectiveness of MT receptor agonists or melatonin in treating depression. Moreover, these data suggest that for the treatment of depression, MT receptor agonists should be targeted directly at the MT1 receptor (Wu et al. 2013). Various treatments for depression, including as bright light therapy, wake therapy, social rhythm therapy, and certain antidepressant drugs (such as SSRIs, SNRIs, and agomelatine), directly influence the body's circadian rhythms (Germain and Kupfer 2008). Implementing morning bright light therapy results in a modification of the melatonin rhythm, and the magnitude of this modification is directly linked to the amelioration of symptoms associated with depression (Terman et al. 2001). In addition, the use of antidepressant medications such as SSRIs, SNIRs, and agomelatine (which acts as an MT1 and MT2 agonist and a 5HT2c antagonist) leads to an earlier shift in the timing of circadian rhythms, including melatonin, body temperature, activity, and cortisol cycles. Wake therapy directly alters sleep-wake cycles by increasing the amount of slow-wave sleep (SWS), reducing the time it takes to enter rapid eye movement (REM) sleep, and shortening the duration of REM sleep (van Berger et al. 2003). Therefore, the successful administration of chronotherapies for Major Depressive Disorder (MDD), in conjunction with the documented variations in symptoms over the course of the day, provide evidence to support the theory that disruptions in the body's natural sleep-wake cycle may play a significant role in the onset of this condition.

7.4 Hypertension

Hypertension is the predominant risk factor and is accountable for the greatest number of fatalities across all causes. The global prevalence of hypertension is around 1.4 billion, representing approximately 31% of the adult population globally (Mills et al. 2020). Diagnosis typically depends on blood pressure (BP) values taken at a medical office or clinic, which do not consider variations in the daily rhythm of BP. Circadian rhythms are 24-h oscillations that play a crucial role in coordinating several physiological systems, such as blood pressure management (as discussed in Douma and Gumz 2018). The daily blood pressure pattern is characterised by a rapid surge in the morning upon waking up, which is then followed by a consistent level throughout the day and a slow decline throughout the night (resulting in a fall of 10–20% from the average blood pressure throughout the day). Ambulatory blood pressure monitoring is essential for effectively assessing the variations in blood pressure across a 24-h period (Mancia et al. 2013). Emerging research indicates that circadian clocks have a substantial impact on the regulation of most cardiovascular functional changes. BP, which plays a crucial role in cardiovascular health, exhibits a clear diurnal pattern characterised by a reduction during the night and a spike in the morning in both humans and mice (Smolensky et al. 2017). Several variables can impact the circadian rhythm of blood pressure. For instance, the hormone melatonin, secreted by the pineal gland during the night, can significantly reduce nocturnal blood pressure and improve sleep quality in persons with severe hypertension (Scheer et al. 2004). The periodicity of atrial natriuretic peptide (ANP) was shown to be in opposite phase with the blood pressure (BP) rhythm, which has been shown to be a vital regulating factor in the 24-h BP pattern and its influence on the cardiovascular system (Portaluppi et al. 1997). In addition, the 24-h blood pressure rhythm is also influenced by the circadian rhythms of plasma renin activity (PRA), angiotensin-converting enzyme (ACE) activity, concentration of angiotensin II (AngII), aldosterone, and thyroid hormone. In addition to hormones, many elements of blood arteries, such as smooth muscle cells and endothelial cells, together with several vasoactive substances, possess the capability to modify the daily variations in blood pressure. Endothelial cells commonly secrete nitric oxide (NO) to control vascular tone, hence impacting blood pressure (BP). Nitric oxide (NO) production is increased in the morning, which is then accompanied by a sudden rise in blood pressure (BP) during this time of day. Nevertheless, there is a current dispute over the direct relationship between the daily variation of NO levels and the rhythmicity of blood pressure. The disruption of the daily variation in the production of nitric oxide has been substantially associated with alterations in blood pressure in cardiovascular disease (Böger et al. 2000). Endothelial NO synthase (eNOS) is one of the three enzymes that play a role in producing nitric oxide (NO) in blood arteries, which in turn regulates vascular function. During animal tests, it was observed that the phosphorylated-eNOS (p-eNOS) in the blood arteries of young mice exhibited a circadian rhythm. The central clock genes can regulate the interaction between eNOS and aid in maintaining the regularity of endothelial function and blood pressure (Anea et al. 2012). In addition, previous research has found a consistent pattern of a 3.5-day cycle and 8-hourly fluctuations in human circulation. Moreover, it has been firmly established that the release of endothelin 1 (ET-1) in the bloodstream and its elimination through urine exhibit a daily rhythm and are closely connected to the regulation of blood pressure. ET-1 receptor antagonists have been shown to be effective in treating essential hypertension. However, significant adverse effects such as fluid retention and edoema have also been reported (Rautureau and Schiffrin 2012). Aside from the aforementioned variables, there are various additional internal mechanisms that contribute to the 24-h oscillations in blood pressure via the utilisation of chemical, hormonal, neurological, or other coordinated regulatory signals. Furthermore, certain illnesses have the capacity to alter the circadian blood pressure rhythm. Individuals diagnosed with obstructive sleep apnea (OSA) are more prone to having non-dipper blood pressure (BP), which in turn increases the risk of cardiovascular events (Cuspidi et al. 2019). Research has shown that sleep habits can significantly affect the day-night blood pressure profile, which is an exogenous factor (Kario et al. 2003). Patients who experience sleep disturbance or circadian misalignment due to shift work and social jet lag often have hypertension (Riegel et al. 2019), delayed blood pressure phase (Motohashi et al. 1998), irregular melatonin secretion rhythm (Gumenyuk et al. 2012), or elevated levels of high-sensitivity C-reactive protein (Morris et al. 2017). Furthermore, research has demonstrated that the daily variations in blood pressure are mostly driven by the internal biological clock rather than by behaviour, particularly under typical circumstances (Yang et al. 2016). Table 7.3 presents a comprehensive explanation of various clock genes and their function in the regulation of hypertension.

To fully understand the impact of the circadian system on hypertension, it is necessary to do future research to explore additional risk factors that are associated with the control of blood pressure. Myocardial infarction (MI), which is another risk factor for blood pressure issues, is a severe cardiac disorder that can be exacerbated by arrhythmia, shock, or heart failure and poses a significant risk to life. There is an unavoidable correlation between the rhythm of blood pressure (BP) and myocardial infarction (MI). Deviant circadian patterns of blood pressure or long-term high blood pressure can result in alterations in the functioning of the heart and raise the likelihood of a myocardial infarction (Virag and Lust 2014). Several clinical research has shown that myocardial infarction (MI) exhibits a rhythmic pattern, with a significantly greater occurrence rate often observed between 06:00 am and

Table 7.3 The impact of manipulating clock gene(s) on blood pressure in rodents

Species	Gene KO in whole organism	Effects on BP	Circadian rhythm of BP
Mice	Bmal1	Decrease	Not present
	Clock	Decrease	Present
	Per1 (129/sv)	Decrease	Present
	Per1	Increase	Present
	Per2	No change	Present
	Cry1/2	Salt sensitivity	Not present
Rat	Bmal1	Decrease	Present

12:00 pm. This discovery is consistent with the increase in blood pressure that occurs in the morning. In the post-myocardial infarction phase, having a systolic blood pressure below 100 mmHg increases the risk of cardiovascular events. On the other hand, those with a systolic blood pressure above 140 mm Hg have a higher risk of stroke and combined cardiovascular events (Thune et al. 2008). Resting systolic blood pressure is less effective than systolic blood pressure following exercise in predicting the occurrence of acute myocardial infarction (Laukkanen et al. 2004). Individuals who have a diastolic blood pressure below 70 mmHg and undergo acute myocardial infarction (AMI) are at increased risk following surgery. The risks encompass mortality from any cause, mortality associated with cardiovascular complications, and hospitalisation resulting from cardiovascular disorders. In addition, wide pulse pressure was also considered a predictive sign of myocardial infarction (Safar 2009). In addition, certain fundamental clock genes are regarded as essential elements in controlling the rhythmic development of myocardial infarction. Škrlec et al. (2020) have demonstrated a link between myocardial infarction (MI) and gene polymorphisms of Clock and Bmal1. In addition, genetic variations of the Arntl, Clock, and Per2 genes have been demonstrated to be associated with circadian characteristics (such as chronotype or daytime drowsiness) in patients with myocardial infarction.

In the context of heart failure with intact ejection fraction, the heightened fluctuation in both systolic and diastolic blood pressure is associated with an elevated risk of negative outcomes (Wei et al. 2020). Simultaneously, inaccurate systolic blood pressure (BP) and diastolic BP might cause fluctuations in cardiac symptoms. According to Wei et al. (2020), an increase in systolic blood pressure is associated with an increase in the thickness of the left ventricular wall and the left ventricular mass index. On the other hand, the functional index of left ventricular diastolic function is inversely related to an increase in diastolic blood pressure, but it is not affected by systolic blood pressure. Typically, there is a U-shaped relationship between systolic blood pressure (BP) and heart failure. Both lower and higher systolic blood pressure readings are correlated with an elevated risk of mortality and hospitalisation caused by heart failure. Therefore, in order to precisely diagnose myocardial infarction (MI) and other cardiovascular events, it is crucial to thoroughly assess the characteristics and patterns of blood pressure (BP). In addition to the influence of clock genes, researchers have also examined the direct impact of the SCN on blood pressure regulation. An investigation carried out following the demise of primary hypertension patients who suffered from myocardial infarction or brain haemorrhage, in comparison to individuals with normal blood pressure, discovered that the immunoreactivity staining for the three primary neuronal populations of the SCN (AVP, VIP, and neurotensin) was diminished by more than 50% in hypertension patients as opposed to the control group (Goncharuk et al. 2001).

A significant increase in the expression of CRH was observed in all areas of the hypothalamic PVN in patients diagnosed with hypertension. In addition, the hypertension patients showed a notable increase in the number of CRH fibres that extended from the front part of the paraventricular nucleus (PVN) to the median eminence. These fibres then innervated the back section of the suprachiasmatic nucleus (SCN). Quantitative analysis revealed that the area occupied by CRH fibres in the SCN of patients with hypertension was approximately three times larger than that in the SCN of control subjects. In addition, a distinct inverse correlation was discovered between the size of CRH fibres and the quantity of AVP- or neurotensin-producing neurons within the SCN (Goncharuk 2021). These findings demonstrate a significant disruption of the SCN in persons with hypertension. A stoppage like this might cause a harmful disturbance in the blood flow, which can negatively affect circulation, especially in the morning when there is a change in the balance between being inactive and active. The difficulty of shifting from a state of inactivity to one of physical activity may add to the incidence of cardiovascular events in the morning. Enhancing therapeutic strategies based on the specific characteristics of circadian rhythm modifications is crucial due to the significant link between acute cardiovascular events and blood pressure cycles. Furthermore, it has been noted that the timing of cardiovascular surgery can impact tissue healing and the outcomes of cardiovascular diseases, in addition to optimising drug administration timing. However, a deeper understanding of the connection between the circadian clock and cardiovascular illness could aid in the development of more specific and customised therapy approaches for patients.

7.5 Summary

Circadian rhythms are an essential component of physiology to keep one's health in the best possible condition. One of the unfortunate consequences of the natural process of ageing is that it causes a decrease in the regularity of circadian rhythms, which in turn hinders the generation of melatonin. Circadian disruption is responsible for the generation of age-dependent brain deficiencies in human and animal models. It also results in a wide range of physiological abnormalities, which ultimately contribute to a decrease in quality of life. Furthermore, in contemporary lifestyles, the tolerance for excessive use of light during the nighttime hours further jeopardises the operation of the circadian core clock mechanism, particularly in elderly persons. We are able to conclude that there is a potential two-way connection between circadian disturbance and the start of neurodegenerative disorders. This is because the circadian rhythm is not only disturbed in neurological illnesses such as Alzheimer's disease, Parkinson's disease, and Huntington's disease, but it also plays a role in the development of these diseases.

According to the information presented in the chapter, substantial evidence suggests that people afflicted with neurodegeneration and other conditions associated with the circadian rhythm experience disruptions in their sleep and wake cycle. The disruption of the circadian system playing a substantial role in the occurrence of these symptoms is extremely possible and should be considered a contributing factor. The circadian disruption that has been documented in humans can be reproduced in mouse and Drosophila models, which in turn gives the possibility of doing mechanistic research. It is of the utmost importance to test the hypothesis that circadian disruption is a key component of the disease, and medicines that are based on circadian rhythms play an important part in this process. In an ideal scenario, we would like to prevent the sleep-wake cycle from becoming more disrupted. In the best-case scenario, we might be able to delay the onset of symptoms by improving the regulation of proteostasis or by reducing inflammation through the use of circadian control. It is possible that it would be able to enhance the quality of life of patients who are suffering from neurodegenerative diseases without affecting the fundamental pathophysiology of the degenerative condition. Due to the great variety in the alterations of the circadian system, this particular group presents a significant challenge that needs to be addressed. When it comes to the management of neurodevelopmental disorders (NDD), the implementation of a planned regimen featuring light exposure, meal timing, and physical activity has the potential to be a method that is both cost-effective and efficient. The stratification of participants in clinical trials to concentrate on those most likely to benefit from a specific treatment is likely to be the most important factor in determining whether or not the trials will be successful.

References

Andrich J, Schmitz T, Saft C, Postert T, Kraus P, Epplen J, Przuntek H, Agelink M (2002) Autonomic nervous system function in Huntington's disease. J Neurol Neurosurg Psychiatry 72(6):726–731

Anea CB, Cheng B, Sharma S, Kumar S, Caldwell RW, Yao L, Ali MI, Merloiu AM, Stepp DW, Black SM (2012) Increased superoxide and endothelial NO synthase uncoupling in blood vessels of Bmal1-knockout mice. Circ Res 111(9):1157–1165

Archer SN, Laing EE, Möller-Levet CS, van der Veen DR, Bucca G, Lazar AS, Santhi N, Slak A, Kabiljo R, von Schantz M (2014) Mistimed sleep disrupts circadian regulation of the human transcriptome. Proc Natl Acad Sci 111(6):E682–E691

Arnulf I, Nielsen J, Lohmann E, Schieffer J, Wild E, Jennum P, Konofal E, Walker M, Oudiette D, Tabrizi S (2008) Rapid eye movement sleep disturbances in Huntington disease. Arch Neurol 65(4):482–488

Aziz NA, Jurgens CK, Landwehrmeyer GB, EHDN Registry Study Group, van Roon-Mom W, van Ommen G, Stijnen T, Roos R (2009a) Normal and mutant HTT interact to affect clinical severity and progression in Huntington disease. Neurology 73(16):1280–1285

Aziz NA, Pijl H, Frölich M, Schröder-van der Elst JP, van der Bent C, Roelfsema F, Roos RA (2009b) Delayed onset of the diurnal melatonin rise in patients with Huntington's disease. J Neurol 256:1961–1965

Belanoff JK, Kalehzan M, Sund B, Fleming Ficek SK, Schatzberg AF (2001) Cortisol activity and cognitive changes in psychotic major depression. Am J Psychiatry 158(10):1612–1616

Belmaker RH, Agam G (2008) Major depressive disorder. N Engl J Med 358(1):55–68

Berdyński M, Miszta P, Safranow K, Andersen PM, Morita M, Filipek S, Żekanowski C, Kuźma-Kozakiewicz M (2022) SOD1 mutations associated with amyotrophic lateral sclerosis analysis of variant severity. Sci Rep 12(1):103

Björkqvist M, Wild EJ, Thiele J, Silvestroni A, Andre R, Lahiri N, Raibon E, Lee RV, Benn CL, Soulet D (2008) A novel pathogenic pathway of immune activation detectable before clinical onset in Huntington's disease. J Exp Med 205(8):1869–1877

Boentert M (2020) Sleep and sleep disruption in amyotrophic lateral sclerosis. Curr Neurol Neurosci Rep 20:1–8

Boeve BF (2013) Idiopathic REM sleep behaviour disorder in the development of Parkinson's disease. Lancet Neurol 12(5):469–482

Böger RH, Bode-Böger SM, Tsao PS, Lin PS, Chan JR, Cooke JP (2000) An endogenous inhibitor of nitric oxide synthase regulates endothelial adhesiveness for monocytes. J Am Coll Cardiol 36(7):2287–2295

Bolitho SJ, Naismith SL, Rajaratnam SM, Grunstein RR, Hodges JR, Terpening Z, Rogers N, Lewis SJG (2014) Disturbances in melatonin secretion and circadian sleep–wake regulation in Parkinson disease. Sleep Med 15(3):342–347

Borgers AJ, Bisschop PH, Swaab DF, Fliers E, Alkemade A (2011) Distribution of serotonin transporters in the human hypothalamus. In: Hypothalamic functions in patients with pituitary insufficiency, vol 99

Braun MC, Castillo-Ruiz A, Indic P, Jung DY, Kim JK, Brown RH Jr, Swoap SJ, Schwartz WJ (2019) Defective daily temperature regulation in a mouse model of amyotrophic lateral sclerosis. Exp Neurol 311:305–312

Breen DP, Vuono R, Nawarathna U, Fisher K, Shneerson JM, Reddy AB, Barker RA (2014) Sleep and circadian rhythm regulation in early Parkinson disease. JAMA Neurol 71(5):589–595

Browne SE (2008) Mitochondria and Huntington's disease pathogenesis: insight from genetic and chemical models. Ann N Y Acad Sci 1147(1):358–382

Brzezinski A, Rai S, Purohit A, Pandi-Perumal SR (2021) Melatonin, clock genes, and mammalian reproduction: what is the link? Int J Mol Sci 22(24):13240

Buysse DJ (2005) Diagnosis and assessment of sleep and circadian rhythm disorders. J Psychiatr Pract 11(2):102–115

Cermakian N, Boivin DB (2003) A molecular perspective of human circadian rhythm disorders. Brain Res Rev 42(3):204–220

Cuspidi C, Tadic M, Sala C, Gherbesi E, Grassi G, Mancia G (2019) Blood pressure non-dipping and obstructive sleep apnea syndrome: a meta-analysis. J Clin Med 8(9):1367

Dibner C, Schibler U, Albrecht U (2010) The mammalian circadian timing system: organization and coordination of central and peripheral clocks. Annu Rev Physiol 72:517–549

Dijk D-J, Czeisler CA (1995) Contribution of the circadian pacemaker and the sleep homeostat to sleep propensity, sleep structure, electroencephalographic slow waves, and sleep spindle activity in humans. J Neurosci 15(5):3526–3538

Dinges DF, Pack F, Williams K, Gillen KA, Powell JW, Ott GE, Aptowicz C, Pack AI (1997) Cumulative sleepiness, mood disturbance, and psychomotor vigilance performance decrements during a week of sleep restricted to 4–5 hours per night. Sleep 20(4):267–277

Douma LG, Gumz ML (2018) Circadian clock-mediated regulation of blood pressure. Free Radic Biol Med 119:108–114

Du Z, Tertrais M, Courtand G, Leste-Lasserre T, Cardoit L, Masmejean F, Halgand C, Cho YH, Garret M (2017) Differential alteration in expression of striatal GABAAR subunits in mouse models of Huntington's disease. Front Mol Neurosci 10:198

Emens J, Lewy A, Kinzie JM, Arntz D, Rough J (2009) Circadian misalignment in major depressive disorder. Psychiatry Res 168(3):259–261

Fahrenkrug J, Popovic N, Georg B, Brundin P, Hannibal J (2007) Decreased VIP and VPAC 2 receptor expression in the biological clock of the R6/2 Huntington's disease mouse. J Mol Neurosci 31:139–148

Fifel K, Videnovic A (2019) Chronotherapies for Parkinson's disease. Prog Neurobiol 174:16–27

Gabery S, Georgiou-Karistianis N, Lundh SH, Cheong RY, Churchyard A, Chua P, Stout JC, Egan GF, Kirik D, Petersén Å (2015) Volumetric analysis of the hypothalamus in Huntington disease using 3T MRI: the IMAGE-HD study. PLoS One 10(2):e0117593

Garcia-Borreguero D, Larrosa O, Granizo JJ, de la Llave Y, Hening WA (2004) Circadian variation in neuroendocrine response to L-dopa in patients with restless legs syndrome. Sleep 27(4):669–673

Germain A, Kupfer DJ (2008) Circadian rhythm disturbances in depression. Hum Psychopharmacol Clin Exp 23(7):571–585

Ghaemi M, Rudolf J, Hilker R, Herholz K, Heiss W (2001) Increased pineal Fdopa uptake is related to severity of Parkinson's disease–a PET study. J Pineal Res 30(4):213–219

Goncharuk VD (2021) The hypothalamus and its role in hypertension. Handb Clin Neurol 182:333–354

Goncharuk VD, Van Heerikhuize J, Dai J, Swaab DF, Buijs RM (2001) Neuropeptide changes in the suprachiasmatic nucleus in primary hypertension indicate functional impairment of the biological clock. J Comp Neurol 431(3):320–330

Gossye H, Engelborghs S, Van Broeckhoven C, van der Zee J (2020) C9orf72 frontotemporal dementia and/or amyotrophic lateral sclerosis. GeneReviews®

Gumenyuk V, Roth T, Drake CL (2012) Circadian phase, sleepiness, and light exposure assessment in night workers with and without shift work disorder. Chronobiol Int 29(7):928–936

Hampp G, Ripperger JA, Houben T, Schmutz I, Blex C, Perreau-Lenz S, Brunk I, Spanagel R, Ahnert-Hilger G, Meijer JH (2008) Regulation of monoamine oxidase a by circadian-clock components implies clock influence on mood. Curr Biol 18(9):678–683

Hardiman O, Al-Chalabi A, Brayne C, Beghi E, van den Berg LH, Chio A, Martin S, Logroscino G, Rooney J (2017) The changing picture of amyotrophic lateral sclerosis: lessons from European registers. J Neurol Neurosurg Psychiatry 88(7):557–563

Harper DG, Stopa EG, Kuo-Leblanc V, McKee AC, Asayama K, Volicer L, Kowall N, Satlin A (2008) Dorsomedial SCN neuronal subpopulations subserve different functions in human dementia. Brain 131(6):1609–1617

Haugh RM, Markesbery WR (1983) Hypothalamic astrocytoma. Syndrome of hyperphagia, obesity, and disturbances of behavior and endocrine and autonomic function. Arch Neurol 40(9):560–563. https://doi.org/10.1001/archneur.1983.04050080060011

Hidaka BH (2012) Depression as a disease of modernity: explanations for increasing prevalence. J Affect Disord 140(3):205–214

Hofman MA, Swaab DF (1994) Alterations in circadian rhythmicity of the vasopressin-producing neurons of the human suprachiasmatic nucleus (SCN) with aging. Brain Res 651(1–2):134–142

Hood S, Amir S (2017) The aging clock: circadian rhythms and later life. J Clin Invest 127(2):437–446

Huang Z, Liu Q, Peng Y, Dai J, Xie Y, Chen W, Long S, Pei Z, Su H, Yao X (2018) Circadian rhythm dysfunction accelerates disease progression in a mouse model with amyotrophic lateral sclerosis. Front Neurol 9:218

Irwin M (2002) Effects of sleep and sleep loss on immunity and cytokines. Brain Behav Immun 16(5):503–512

Joo EY, Kim H, Suh S, Hong SB (2014) Hippocampal substructural vulnerability to sleep disturbance and cognitive impairment in patients with chronic primary insomnia: magnetic resonance imaging morphometry. Sleep 37(7):1189–1198

Kalliolia E, Silajdžić E, Nambron R, Hill NR, Doshi A, Frost C, Watt H, Hindmarsh P, Björkqvist M, Warner TT (2014) Plasma melatonin is reduced in Huntington's disease. Mov Disord 29(12):1511–1515

Kario K, Pickering TG, Umeda Y, Hoshide S, Hoshide Y, Morinari M, Murata M, Kuroda T, Schwartz JE, Shimada K (2003) Morning surge in blood pressure as a predictor of silent and clinical cerebrovascular disease in elderly hypertensives: a prospective study. Circulation 107(10):1401–1406

Killoy KM, Pehar M, Harlan BA, Vargas MR (2021) Altered expression of clock and clock-controlled genes in a hSOD1-linked amyotrophic lateral sclerosis mouse model. FASEB J 35(2):e21343

Kim D, Jeon J, Cheong E, Kim DJ, Ryu H, Seo H, Kim YK (2016) Neuroanatomical visualization of the impaired striatal connectivity in Huntington's disease mouse model. Mol Neurobiol 53:2276–2286

Knapen SE, Riemersma-Van der Lek RF, Antypa N, Meesters Y, Penninx BW, Schoevers RA (2018) Social jetlag and depression status: results obtained from The Netherlands study of depression and anxiety. Chronobiol Int 35(1):1–7

Kobal J, Lovrečič L, Peterlin B (2012) Biomarkers for Huntington's disease. In: Huntington's disease-core concepts and current advances. IntechOpen, London

Kräuchi K, Cajochen C, Werth E, Wirz-Justice A (2002) Alteration of internal circadian phase relationships after morning versus evening carbohydrate-rich meals in humans. J Biol Rhythm 17(4):364–376

Kruijver FP, Swaab DF (2002) Sex hormone receptors are present in the human suprachiasmatic nucleus. Neuroendocrinology 75(5):296–305

Kudo T, Schroeder A, Loh DH, Kuljis D, Jordan MC, Roos KP, Colwell CS (2011) Dysfunctions in circadian behavior and physiology in mouse models of Huntington's disease. Exp Neurol 228(1):80–90

Laukkanen JA, Kurl S, Salonen R, Lakka TA, Rauramaa R, Salonen JT (2004) Systolic blood pressure during recovery from exercise and the risk of acute myocardial infarction in middle-aged men. Hypertension 44(6):820–825

Lazar AS, Panin F, Goodman AO, Lazic SE, Lazar ZI, Mason SL, Rogers L, Murgatroyd PR, Watson LP, Singh P (2015) Sleep deficits but no metabolic deficits in premanifest H untington's disease. Ann Neurol 78(4):630–648

Lee A, Myung S-K, Cho JJ, Jung Y-J, Yoon JL, Kim MY (2017) Night shift work and risk of depression: meta-analysis of observational studies. J Korean Med Sci 32(7):1091

Liu R-Y, Zhou J-N, Hoogendijk WJ, van Heerikhuize J, Kamphorst W, Unmehopa UA, Hofman MA, Swaab DF (2000) Decreased vasopressin gene expression in the biological clock of Alzheimer disease patients with and without depression. J Neuropathol Exp Neurol 59(4):314–322

Liu AC, Welsh DK, Ko CH, Tran HG, Zhang EE, Priest AA, Buhr ED, Singer O, Meeker K, Verma IM (2007) Intercellular coupling confers robustness against mutations in the SCN circadian clock network. Cell 129(3):605–616

Lo JC, Loh KK, Zheng H, Sim SK, Chee MW (2014) Sleep duration and age-related changes in brain structure and cognitive performance. Sleep 37(7):821–821

Mai J, Kedziora O, Teckhaus L, Sofroniew M (1991) Evidence for subdivisions in the human suprachiasmatic nucleus. J Comp Neurol 305(3):508–525

Mamoor S (2022) Differential expression of PLEKHA2 in amyotrophic lateral sclerosis

Mancia G, Bombelli M, Brambilla G, Facchetti R, Sega R, Toso E, Grassi G (2013) Long-term prognostic value of white coat hypertension: an insight from diagnostic use of both ambulatory and home blood pressure measurements. Hypertension 62(1):168–174

Maywood ES, Fraenkel E, McAllister CJ, Wood N, Reddy AB, Hastings MH, Morton AJ (2010) Disruption of peripheral circadian timekeeping in a mouse model of Huntington's disease and its restoration by temporally scheduled feeding. J Neurosci 30(30):10199–10204

McNeely E, Mordukhovich I, Tideman S, Gale S, Coull B (2018) Estimating the health consequences of flight attendant work: comparing flight attendant health to the general population in a cross-sectional study. BMC Public Health 18:1–11

Menza M, Dobkin RD, Marin H, Mark MH, Gara M, Bienfait K, Dicke A, Kusnekov A (2010) The role of inflammatory cytokines in cognition and other non-motor symptoms of Parkinson's disease. Psychosomatics 51(6):474–479

Miller GE, Rohleder N, Stetler C, Kirschbaum C (2005) Clinical depression and regulation of the inflammatory response during acute stress. Psychosom Med 67(5):679–687

Mills KT, Stefanescu A, He J (2020) The global epidemiology of hypertension. Nat Rev Nephrol 16(4):223–237

Mirmiran M, Swaab D, Hofman M, Witting W, Kok J, Van Gool W (1992) Circadian rhythms and the suprachiasmatic nucleus in perinatal development, aging and Alzheimer's disease. Prog Brain Res 93:151–163

Moore RY, Speh JC, Leak RK (2002) Suprachiasmatic nucleus organization. Cell Tissue Res 309(1):89–98. https://doi.org/10.1007/s00441-002-0575-2

Morin LP, Blanchard JH (2001) Neuromodulator content of hamster intergeniculate leaflet neurons and their projection to the suprachiasmatic nucleus or visual midbrain. J Comp Neurol 437(1):79–90

Morris CJ, Purvis TE, Mistretta J, Hu K, Scheer FA (2017) Circadian misalignment increases C-reactive protein and blood pressure in chronic shift workers. J Biol Rhythm 32(2):154–164

Motohashi Y, Higuchi S, Maeda A, Liu Y, Yuasa T, Motohashi K, Nakamura K (1998) Alteration of circadian time structure of blood pressure caused by night shift schedule. Occup Med 48(8):523–528

Nagatsu T, Sawada M (2009) L-dopa therapy for Parkinson's disease: past, present, and future. Parkinsonism Relat Disord 15:S3–S8

Nakagawa H, Okumura N (2010) Coordinated regulation of circadian rhythms and homeostasis by the suprachiasmatic nucleus. Proc Jpn Acad Ser B 86(4):391–409

Neutel D, Tchikviladzé M, Charles P, Leu-Semenescu S, Roze E, Durr A, Arnulf I (2015) Nocturnal agitation in Huntington disease is caused by arousal-related abnormal movements rather than by rapid eye movement sleep behavior disorder. Sleep Med 16(6):754–759

Oei NYL, Everaerd WTAM, Elzinga BM, van Well S, Bermond B (2006) Psychosocial stress impairs working memory at high loads: an association with cortisol levels and memory retrieval. Stress (Amsterdam, Netherlands) 9(3):133–141. https://doi.org/10.1080/10253890600965773

Oenning NSX, Ziegelmann PK, De Goulart BNG, Niedhammer I (2018) Occupational factors associated with major depressive disorder: a Brazilian population-based study. J Affect Disord 240:48–56

Piano C, Losurdo A, Della Marca G, Solito M, Calandra-Buonaura G, Provini F, Bentivoglio AR, Cortelli P (2015) Polysomnographic findings and clinical correlates in Huntington disease: a cross-sectional cohort study. Sleep 38(9):1489–1495

Poewe W, Högl B (2000) Parkinson's disease and sleep. Curr Opin Neurol 13(4):423–426

Politis M, Pavese N, Tai YF, Tabrizi SJ, Barker RA, Piccini P (2008) Hypothalamic involvement in Huntington's disease: an in vivo PET study. Brain 131(11):2860–2869

Pontone GM, Williams JR, Anderson KE, Chase G, Goldstein SR, Grill S, Hirsch ES, Lehmann S, Little JT, Margolis RL (2013) Pharmacologic treatment of anxiety disorders in Parkinson disease. Am J Geriatr Psychiatry 21(6):520–528

Portaluppi F, Provini F, Cortelli P, Plazzi G, Bertozzi N, Manfredini R, Fersini C, Lugaresi E (1997) Undiagnosed sleep-disordered breathing among male nondippers with essential hypertension. J Hypertens 15(11):1227–1233

Rautureau Y, Schiffrin EL (2012) Endothelin in hypertension: an update. Curr Opin Nephrol Hypertens 21(2):128–136

Riegel B, Daus M, Lozano AJ, Malone SK, Patterson F, Hanlon AL (2019) Shift workers have higher blood pressure medicine use, but only when they are short sleepers: a longitudinal UK biobank study. J Am Heart Assoc 8(20):e013269

Riemersma-van Der Lek RF, Swaab DF, Twisk J, Hol EM, Hoogendijk WJ, Van Someren EJ (2008) Effect of bright light and melatonin on cognitive and noncognitive function in elderly residents of group care facilities: a randomized controlled trial. JAMA 299(22):2642–2655

Roenneberg T, Merrow M (2016) The circadian clock and human health. Curr Biol 26(10):R432–R443

Romijn H, Van Uum J, Emmering J, Goncharuk V, Buijs R (1999) Colocalization of VIP with AVP in neurons of the human paraventricular, supraoptic and suprachiasmatic nucleus. Brain Res 832(1–2):47–53

Rusting CL, Larsen RJ (1998) Diurnal patterns of unpleasant mood: associations with neuroticism, depression, and anxiety. J Pers 66(1):85–103

Safar ME (2009) Hypertension, systolic blood pressure, and large arteries. Med Clin North Am 93(3):605–619

Sakamoto K, Liu C, Kasamatsu M, Pozdeyev NV, Iuvone PM, Tosini G (2005) Dopamine regulates melanopsin mRNA expression in intrinsically photosensitive retinal ganglion cells. Eur J Neurosci 22(12):3129–3136

Scheer FA, Van Montfrans GA, van Someren EJ, Mairuhu G, Buijs RM (2004) Daily nighttime melatonin reduces blood pressure in male patients with essential hypertension. Hypertension 43(2):192–197

Schmid SM, Hallschmid M, Schultes B (2015) The metabolic burden of sleep loss. Lancet Diabet Endocrinol 3(1):52–62

Schulte EC, Winkelmann J (2011) When Parkinson's disease patients go to sleep: specific sleep disturbances related to Parkinson's disease. J Neurol 258:328–335

Schwartz WJ, Gross RA, Morton MT (1987) The suprachiasmatic nuclei contain a tetrodotoxin-resistant circadian pacemaker. Proc Natl Acad Sci 84(6):1694–1698

Škrlec I, Milić J, Steiner R (2020) The impact of the circadian genes CLOCK and ARNTL on myocardial infarction. J Clin Med 9(2):484

Smolensky MH, Hermida RC, Portaluppi F (2017) Circadian mechanisms of 24-hour blood pressure regulation and patterning. Sleep Med Rev 33:4–16

Soneson C, Fontes M, Zhou Y, Denisov V, Paulsen JS, Kirik D, Petersén Å, Huntington Study Group PREDICT-HD investigators (2010) Early changes in the hypothalamic region in prodromal Huntington disease revealed by MRI analysis. Neurobiol Dis 40(3):531–543

Stopa EG, Volicer L, Kuo-Leblanc V, Harper D, Lathi D, Tate B, Satlin A (1999) Pathologic evaluation of the human suprachiasmatic nucleus in severe dementia. J Neuropathol Exp Neurol 58(1):29–39

Swaab D (2003) Suprachiasmatic nucleus (SCN) and pineal gland. In: Handbook of clinical neurology. The human hypothalamus: basic and clinical aspects—part I: Nuclei of the human hypothalamus, vol 79. Elsevier, Amsterdam, pp 63–125

Swaab D (2004) Neurodegenerative disorders. In: Aminoff MJ, Boller F, Swaab DF (eds) The human hypothalamus: basic and clinical aspects, part II: neuropathology of the human hypothalamus and adjacent structures. Handbook of clinical neurology, vol 80. Elsevier, Amsterdam, pp 313–349

Tabrizi SJ, Flower MD, Ross CA, Wild EJ (2020) Huntington disease: new insights into molecular pathogenesis and therapeutic opportunities. Nat Rev Neurol 16(10):529–546

Terman JS, Terman M, Lo E-S, Cooper TB (2001) Circadian time of morning light administration and therapeutic response in winter depression. Arch Gen Psychiatry 58(1):69–75

Thenganatt MA, Jankovic J (2016) The relationship between essential tremor and Parkinson's disease. Parkinsonism Relat Disord 22:S162–S165

Thune JJ, Signorovitch J, Kober L, Velazquez EJ, McMurray JJ, Califf RM, Maggioni AP, Rouleau JL, Howlett J, Zelenkofske S (2008) Effect of antecedent hypertension and follow-up blood pressure on outcomes after high-risk myocardial infarction. Hypertension 51(1):48–54

Vadnie CA, McClung CA (2017) Circadian rhythm disturbances in mood disorders: insights into the role of the suprachiasmatic nucleus. Neural Plast 2017:1

van Berger M, Van Calker D, Riemann D (2003) Sleep and manipulations of the sleep–wake rhythm in depression. Acta Psychiatr Scand 108:83–91

Van de Nes J, Kamphorst W, Ravid R, Swaab D (1998) Comparison of β-protein/A4 deposits and Alz-50-stained cytoskeletal changes in the hypothalamus and adjoining areas of Alzheimer's disease patients: Amorphic plaques and cytoskeletal changes occur independently. Acta Neuropathol 96:129–138

van Londen L, Goekoop JG, Kerkhof GA, Zwinderman KH, Wiegant VM, De Wied D (2001) Weak 24-h periodicity of body temperature and increased plasma vasopressin in melancholic depression. Eur Neuropsychopharmacol 11(1):7–14

Van Someren EJ, Raymann RJ, Scherder EJ, Daanen HA, Swaab DF (2002) Circadian and age-related modulation of thermoreception and temperature regulation: mechanisms and functional implications. Ageing Res Rev 1(4):721–778

van Wamelen DJ, Martinez-Martin P, Weintraub D, Schrag A, Antonini A, Falup-Pecurariu C, Odin P, Ray Chaudhuri K, International Parkinson and Movement Disorder Society Parkinson's Disease Non-Motor Study Group (2021) The non-motor symptoms scale in Parkinson's disease: validation and use. Acta Neurol Scand 143(1):3–12

Videnovic A, Lazar AS, Barker RA, Overeem S (2014) The clocks that time us'—circadian rhythms in neurodegenerative disorders. Nat Rev Neurology 10(12):683–693

Virag JA, Lust RM (2014) Circadian influences on myocardial infarction. Front Physiol 5:116760

Waddington Lamont E, Legault-Coutu D, Cermakian N, Boivin DB (2007) The role of circadian clock genes in mental disorders. Dialogues Clin Neurosci 9(3):333–342

Wei J, Galaviz KI, Kowalski AJ, Magee MJ, Haw JS, Narayan KV, Ali MK (2020) Comparison of cardiovascular events among users of different classes of antihypertension medications: a systematic review and network meta-analysis. JAMA Netw Open 3(2):e1921618–e1921618

Wetter T, Collado-Seidel V, Pollmacher T, Yassouridis A, Trenkwalder C (2000) Sleep and periodic leg movement patterns in drug-free patients with Parkinson's disease and multiple system atrophy. Sleep 23(3):361–368

Williams-Gray C, Foltynie T, Brayne C, Robbins T, Barker R (2007) Evolution of cognitive dysfunction in an incident Parkinson's disease cohort. Brain 130(7):1787–1798

Willison LD, Kudo T, Loh DH, Kuljis D, Colwell CS (2013) Circadian dysfunction may be a key component of the non-motor symptoms of Parkinson's disease: insights from a transgenic mouse model. Exp Neurol 243:57–66

Wu Y, Fischer DF, Kalsbeek A, Garidou-Boof M, van der Vliet J, van Heijningen C, Liu R, Zhou J, Swaab DF, Wu Y (2006) Pineal clock gene oscillation is disturbed in Alzheimer's disease, due to functional disconnection from the "master clock". FASEB J 20(11):1874–1876

Wu Y-H, Ursinus J, Zhou J-N, Scheer FA, Ai-Min B, Jockers R, van Heerikhuize J, Swaab DF (2013) Alterations of melatonin receptors MT1 and MT2 in the hypothalamic suprachiasmatic nucleus during depression. J Affect Disord 148(2–3):357–367

Yang G, Chen L, Grant GR, Paschos G, Song W-L, Musiek ES, Lee V, McLoughlin SC, Grosser T, Cotsarelis G (2016) Timing of expression of the core clock gene Bmal1 influences its effects on aging and survival. Sci Transl Med 8(324):324ra16-324ra16

Zhang G, Zhang L, Huang T, Fan D (2020) Daytime sleepiness, night sleep changes and ALS: a Mendelian randomization study

Zhong X, Hilton HJ, Gates GJ, Jelic S, Stern Y, Bartels MN, DeMeersman RE, Basner RC (2005) Increased sympathetic and decreased parasympathetic cardiovascular modulation in normal humans with acute sleep deprivation. J Appl Physiol 98(6):2024–2032

Zhou A-W, Guo J, Wang H-Y, Gu B-X, Du Y-C (1995) Enhancement of NGF gene expression in rat brain by the memory-enhancing peptide AVP (4–8). Peptides 16(4):581–586

Ziemssen T, Reichmann H (2010) Cardiovascular autonomic dysfunction in Parkinson's disease. J Neurol Sci 289(1–2):74–80

The Circadian Clock as a Drug Target for Anti-Ageing Strategies

8

8.1 Introduction

Aging is commonly perceived as a gradual and continuous deterioration of bodily functions that occurs as one gets older. This decrease is often linked to a reduced ability to handle injury, sickness, and stress (Poon and Perls 2007). A broader interpretation of aging, with an evolutionary perspective, is the ongoing decrease in age-specific "fitness components" of an organism as a result of internal physiological deterioration (Rose 1994). Rose has highlighted that all definitions have a tendency to influence our perception of a situation. For instance, both definitions tend to emphasize the adverse aspects of aging. The process of aging can also encompass adaptability. The process of adaptation to the environment can lead to successful survival when an organism acquires knowledge and skills (Martin 2011).

The correlation between the clock and aging is pertinent to several theories of aging. Aging may be considered a temporal occurrence, similar to the timing of events such as salmon migration or puberty. Therefore, the concept of a clock could be closely linked to the aging process, as it is a component of the planned aging mechanism (Witkowski and Hagberg 2007). Programmed aging can encompass apoptosis, which refers to the intentional death of cells (Cohen et al. 2013). The circadian clock is responsible for regulating daily fluctuations in a significant number of genes, comprising a substantial portion of the entire transcriptome (Koike et al. 2012). Consequently, any disruptions in the circadian clock mechanism can impact both physiology and behaviour. Various diseases, including malignancies, metabolic problems, heart disease, sleep disorders, and certain types of mental illness, have been associated with changes or disturbances in circadian rhythms (Jagannath et al. 2017). Moreover, it is widely recognized that the biological clock and aging are regulated by interconnected genetic networks, each comprising numerous genes. Given the wide-ranging influence of circadian rhythms on physiology, there is a keen interest in developing pharmaceuticals that specifically target the circadian clock. Nevertheless, the selection of targets and the specific small

S. I. Rizvi, G. Majumdar, *Circadian Clock and Aging*,
https://doi.org/10.1007/978-981-97-6695-6_8

molecule effectors might lead to various modifications of the clock, necessitating tactics contingent upon the intended result. For instance, enhancing the magnitude of the circadian rhythm could be advantageous for promoting better health in aging.

Conversely, for sleep disorders caused by an elongated circadian rhythm, medications that advance the internal clock would be of interest. Pharmacologically stimulating the circadian clock is seen as a potential approach for anti-ageing treatment. For instance, certain small compounds that specifically affect the circadian clock can also control circadian rhythms, metabolism, cell cycle and/or cell proliferation. The development of these strategies is facilitated by the improved comprehension of the circadian clock mechanism. This understanding has been significantly advanced in recent years through the discovery of atomic-level structures of several crucial components (Xing et al. 2013). In this chapter, we examine the recent discoveries on drugs that can modulate the clock and analyses the biochemical characteristics of the core clock that could potentially be targeted in future therapeutic development related to aging or old age.

8.2 Targeting Clock for the Development of Therapeutics

The circadian clock governs the daily fluctuations of many genes, including a significant portion of the overall transcriptome (Koike et al. 2012; Zhang et al. 2014). Consequently, any disruptions in the circadian clock system can impact physiology and behaviour. Consequently, various diseases have been associated with changes or disturbances in circadian rhythms, including malignancies, metabolic disorders, heart disease, sleep disorders, and multiple forms of mental illness (Allada and Bass 2021; Jagannath et al. 2017). In addition, the process of ageing has been associated with a decrease in the magnitude of the circadian rhythm (Acosta-Rodríguez et al. 2021). Furthermore, as initially demonstrated in the 1970s (Haus et al. 1972), numerous medications designed to treat severe illnesses exhibit varying levels of effectiveness depending on the specific time of day they are administered. A study was conducted using RNA-sequencing (RNA-seq) and DNA arrays to investigate the circadian oscillations in transcript expression in 12 mouse organs. The results revealed that over 50% of the most popular medications target proteins derived from genes that exhibit rhythmic activity (Zhang et al. 2014). The significance of circadian timing in medicine has been progressively illustrated and developed as a distinct area of study referred to as circadian medicine (for details see, (Sancar and Van Gelder 2021)).

Given the extensive influence of circadian rhythms on physiology, there is a keen interest in the development of pharmaceuticals that specifically target the circadian clock. Nevertheless, the selection of targets and the specific small molecule effectors might lead to various modifications of the clock, necessitating tactics contingent upon the intended result. Enhancing the magnitude of the circadian rhythm could be advantageous for promoting better health in the aging process. Conversely, for sleep disorders caused by an elongated circadian rhythm, medications that advance the internal clock would be of interest. Additionally, inhibiting the circadian clock may halt the proliferation of cancer stem cells in certain types of cancers, such as glioblastoma. The development of these strategies is facilitated by the improved comprehension of the circadian clock mechanism, which has been significantly advanced in recent years through the discovery of numerous crucial elements (Fribourgh et al. 2020), as well as through extensive investigation of the components and proportions of the activator and repressor complexes, and their regulatory mechanisms (Fribourgh et al. 2020; Rosensweig and Green 2020). Comprehending the molecular constituents involved in the Transcription-translational feedback loop (TTFL) of the clock can enhance our understanding of how a medicinal substance can achieve the intended outcome. The repression phase begins with an initial repressive stage at Circadian Time 16 (CT16), during which CLOCK: BMAL1: CRY1/2:PER1/2 (and with related proteins) are attached to DNA. This is followed by the poised phase at CT24, where the complex is composed of CLOCK: BMAL1: CRY1. After the degradation of CRY1, the inhibition is lifted, enabling the transcription phase to begin (Koike et al. 2012).

Suppressing the proteins in each of the three states will result in distinct consequences (Rosensweig and Green 2020). By suppressing CRY2, the duration of the transcriptional phase is prolonged, resulting in an extended period of transcription for the circadian-regulated genes in this simplified scenario. Similarly, the inhibition of CRY1 will prolong the transcriptional phase by predominantly reducing the poised phase and enabling an early initiation of the transcriptional phase, resulting in a distinctly different consequence. These findings align with the natural cycle duration observed in CRY knockout mice. Specifically, CRY1 knockout mice have a period of 1.2 h shorter, whereas CRY2 knockout mice have a period of 0.8 h longer (Vitaterna et al. 1999). Demonstrating that despite both proteins being part of the negative arm of the feedback loop, blocking them does not yield identical consequences. This emphasizes the significance of comprehending the specific state of the molecular clock that is being suppressed or amplified when determining which core clock proteins to focus on for therapeutic intervention. The activator proteins of the TTFL, CLOCK and BMAL1 are the most apparent candidates for the regulation of the circadian clock. These proteins have a direct impact on gene expression. If their activity is inhibited by preventing them from binding to DNA, stopping the CLOCK/BMAL1 proteins from forming pairs or blocking their interaction with other positive transcription factors, it is expected that the strength of circadian rhythms would decrease and the gene expression levels of their numerous targets would be reduced. In contrast, drugs that hinder the binding of repressor proteins would enhance the expression of their target genes and perhaps prolong the duration of the active phase. CLOCK and BMAL1 are members of the bHLH-PAS protein family, including several proteins significantly linked to human disease.

The bHLH-PAS transcription factor family derives its name from the unique structural domains that comprise it. The protein consists of an N-terminal basic Helix-Loop-Helix (bHLH) domain, which is then followed by two consecutive PER-ARNT-SIM (PAS) domains. The rest of the protein is mostly unstructured (Wu and Rastinejad 2017). Typically, the bHLH-PAS transcription factor family can be categorised into two classes according to their patterns of heterodimerization. Class

I comprises three Hypoxia-inducible factor (HIF) proteins, namely HIF-1a, HIF-2a, and HIF-3a, neuronal PAS domain proteins (NPAS1–4), aryl hydrocarbon receptor (AHR), AHR repressor (AHRR), single-minded proteins (SIM1–2), and CLOCK. Class II comprises aryl hydrocarbon receptor nuclear translocator (ARNT, ARNT2), BMAL1 (ARNTL), and BMAL2 (ARNTL2). The formation of a heterodimer between a protein belonging to class I and a protein belonging to class II results in the creation of a transcription factor that has the ability to regulate gene expression. As previously stated, a significant number of these proteins have been linked to human disorders. Loss-of-function mutations in the SIM1 gene have been connected to obesity. The SIM2 gene is linked to the development of malignancies. HIF proteins play a role in controlling important genetic processes necessary for tumour regulation. Disruption of the CLOCK and BMAL1 genes can result in a range of metabolic diseases and sleep disorders. These findings are supported by references (Bersten et al. 2013).

8.3 Cellular and Molecular Aspects of Chrono-Pharmacology and Chronotherapy

There is a growing interest in utilizing the clock machinery for health advantages. Classical chronotherapy (Levi and Schibler 2007) aims to determine the best time period for administering a therapy to maximize its effectiveness while minimizing its harmful effects. A significant majority of leading medications, a considerable number of which are designed to address long-lasting illnesses that impact life expectancy, function by targeting protein targets encoded by genes expressed cyclically (Zhang et al. 2014). Together with the circadian regulation of drug metabolism (Levi and Schibler 2007), this discovery suggests that the pharmacokinetic and pharmacodynamics characteristics of popular medications are likely to differ depending on the time of day they are administered. Several disease events or symptoms exhibit a clear circadian preference, such as heart attacks occurring in the early morning and asthma symptoms worsening at night. Timed treatment administration, known as chronotherapies, has been documented for a range of disorders, including cancer (Antoch and Kondratov 2013; Levi and Schibler 2007). Detailed physiological events, their activity peak times and where chrono-based therapies can be applied are depicted in Fig. 8.1. A mechanistic investigation conducted by Gorbacheva and colleagues (Gorbacheva et al. 2005) has established a correlation between the sensitivity to the chemotherapeutic medication cyclophosphamide and the functional status of the key circadian transcription factor CLOCK/BMAL1. The medication was administered to genetic mouse mutants with different levels of CLOCK/BMAL1 (the positive component of the core loop). It was observed that cell toxicity strongly correlated with the expression and/or activity of CLOCK/BMAL1.

An effective approach to take advantage of the regulating function of the clock in disease and aging is to control the clock or its components (Nohara et al. 2015; Wallach and Kramer 2015). Instead of synchronising the timing of therapy with the natural rhythm, one can modify or amplify the natural rhythm to attain the desired

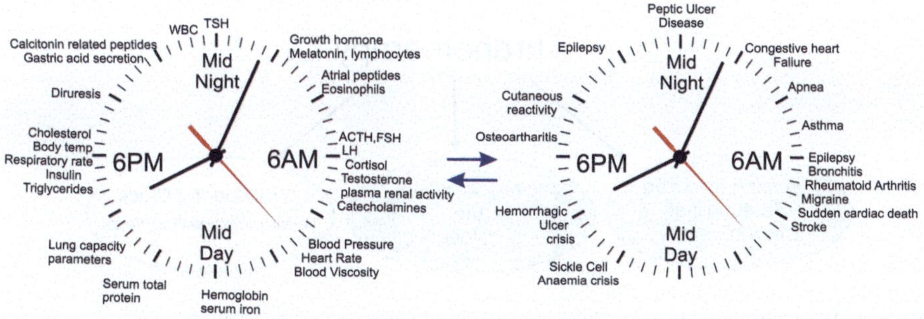

Fig. 8.1 Schematic representation of different physiological events, their activity peak times and their interrelation with health issues which are also time of day dependent

result. Several behavioural and dietary interventions have been identified to modify the circadian rhythm, including exposure to light (Banks et al. 2016), physical exercise (Schroder and Esser 2013), and patterns of eating and fasting (Longo and Panda 2016). Specifically, a recent study demonstrated that time-restricted feeding (TRF) had positive effects on sleep quality, body weight regulation, and the delay of cardiac aging and healthy aging in Drosophila (Gill et al. 2015). Expression study revealed that the activation of genes related to circadian rhythms and the complexes involved in the mitochondrial electron transport chain depends on TRF. CR, like TRF, temporally consolidated food intake, significantly increased the expression and strength of core clock genes, leading to changes in lipid homeostasis and ultimately contributing to its ability to extend lifespan (Katewa et al. 2016; Longo and Panda 2016). Studies conducted in group care facilities have demonstrated the positive impact of strong light and melatonin, which are both significant regulators of circadian rhythms, on the cognitive function and mood of elderly individuals (Riemersma-van Der Lek et al. 2008). In addition, the injection of melatonin into mice, including those with melatonin deficiency (C57BL/6 mice), as well as the implantation of pineal grafts from young to elderly mice, resulted in an extension of longevity (Pierpaoli and Regelson 1994). These studies offer compelling evidence that improving the molecular and physiological rhythms through nutritional and behavioural interventions might enhance metabolism and physiology, leading to healthy aging (Fig. 8.2).

In order to overcome the typical problems associated with behavioural, nutritional, and lifestyle interventions, a promising approach is to use chemical compounds that can manipulate the body's internal clock. This allows for easy and consistent use of these compounds over an extended period of time. Two main methods for finding circadian modulator compounds are unbiased phenotypic screens of huge chemical libraries and tailored approaches focusing on specific clock components (He and Chen 2016). Identification of small molecules through phenotypic screens involves the use of high-throughput circadian reporter experiments, typically utilising Per2 or Bmal1 promoters (Hirota and Kay 2009; Jones et al. 2013). Nevertheless, the identification of protein targets for several drugs has

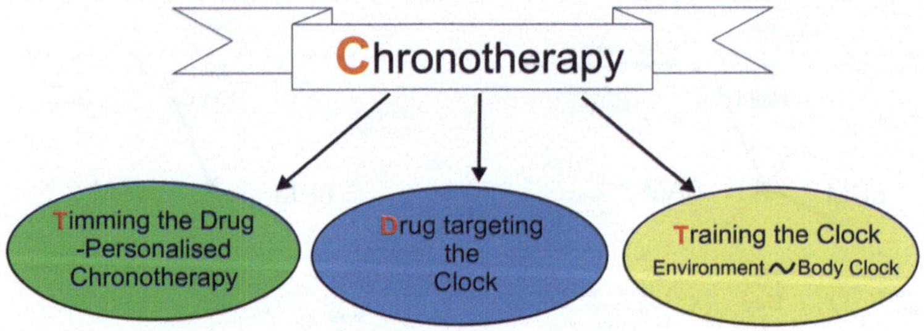

Fig. 8.2 Health based scenarios where chrono-based therapies can be applied

not been achieved through experimental means (He and Chen 2016), posing a significant obstacle to their practical use.

The precise measurement of circadian periodicity has enabled the discovery of numerous substances that have significant impacts on the length of the circadian period. Several small compounds that inhibit casein kinase I (CKI) and have a significant effect on lengthening the circadian periodicity have been found (Hirota et al. 2010; Isojima et al. 2009). CKI plays a primary regulatory role in the circadian rhythm. These results also confirmed the idea that the half-life of the PER protein is a crucial factor in determining the length of the circadian period (Gallego and Virshup 2007). Conventional CKI inhibitors commonly exhibit inhibition of various CKI isoforms, such as CKId and CKIe (Isojima et al. 2009). On the other hand, the drug PF-4800567 was specifically designed to block the CKIe isoform. This finding highlights the significant function of CKId in regulating the biological clock. Recent research has utilised the CKIe-specific ligand to investigate clock-related pathologies, including substance misuse (Bryant et al. 2012). CRY proteins, which form dimers with PERs in the negative part of the core loop, are likewise regulated by complex mechanisms that maintain their stability. Mutations in two similar E3 ligases for CRYs have been identified in mice genetic screens, and these mutations are known to alter the length of the circadian period (Yoo et al. 2013). Biochemical investigations have identified CRYs as the specific protein that KL001 acts upon. The binding of KL001 hinders the process of CRY ubiquitination and the subsequent destruction of CRY. Pilot investigations demonstrated that variants of KL001 enhanced glucose tolerance in mice (Humphries et al. 2016), indicating that activating CRYs may have a metabolic impact.

The REV-ERB and ROR receptor families have opposing roles in the secondary loop of the circadian oscillator. They regulate the transcription of target genes, such as Bmal1, through the RORE pathway (Liu et al. 2007). The multi-functional receptors have a major role in regulating the amplitude of circadian rhythms. Recent research has discovered new mechanisms that regulate amplitude, including loading chromatin and protein turnover after translation (Zhao et al. 2016; Zhu et al. 2015). Several ligands for REV-ERBs and RORs have been documented in recent years (He and Chen 2016; Kojetin and Burris 2014). Multiple investigations have centred on SR9011, which is a REV-ERB agonist created via specific chemical alteration

(Solt et al. 2012). SR9011 altered the expression of clock genes in tissues involved in metabolism and caused a rapid inhibition of circadian wheel-running behaviour.

When diet-induced obese mice were given SR9011, their metabolism, including body weight and levels of glucose and lipids in the blood, improved (Solt et al. 2012). SR9011 has been discovered to enhance alertness in mice treated with it, while simultaneously inhibiting REM and slow-wave sleep (Banerjee et al. 2014). The anxiolytic effect of SR9011 was observed in wild-type mice but not in mice lacking the Rev-erbb gene. The results indicate that REV-ERB agonists have the potential to be used as a therapy to regulate metabolism and behaviour related to the circadian rhythm and sleep.

Several other small compounds that target RORs have been reported in the literature (He and Chen 2016; Helleboid et al. 2014). In addition to its role as a circadian activator in the secondary loop, RORs have been found to play significant roles in the development of the neuronal system, motor function, metabolism, and immunology, as revealed by genetic and genomic investigations (Jetten et al. 2013). The latest breakthroughs have clarified that cholesterol intermediate metabolites could potentially serve as natural ligands (Santori 2015). However, the specific role of this interaction, particularly within the circadian system, remains uncertain. Many studies have focused on inhibitors of RORct and their application in immune response due to their significant role in autoimmune disease (for details, see Huh et al. 2011). Additional research is needed to determine the impact of these RORct ligands on circadian and metabolic functioning.

8.4 Drugs Targeting the Clock

There is a growing interest in focusing on the components of the circadian clock as a new therapeutic method to treat chronic disorders. Chronic disorders like metabolic syndrome, cancer, and chronic inflammatory diseases generally include the disruption of numerous metabolic or signalling pathways. An optimal treatment would aim to either activate or suppress one of these affected processes. By targeting a component of the circadian clock, it is possible to achieve similar advantages as a combination therapy, when many medications target different pathways. Administering medication to the clock components does not guarantee that restoring the daily patterns in the activity of the core circadian clock's activators and repressors is essential for achieving the intended result. The concept of clock manipulation is rooted in the notion that circadian rhythmic outputs preserve homeostasis by temporarily suppressing a pathway, thereby preventing long-term activation that could be harmful. For instance, the circadian cycle provides an inherent mechanism to suppress inflammation, therefore guaranteeing the remission of inflammation without its progression into chronic inflammation. When dealing with long-term inflammation, it is desirable to focus on a clock component that can effectively reduce inflammation. Although a medicine that suppresses inflammation and has a long duration of action may provide advantages, a drug with a short duration of action but administered at the ideal time may restore the daily rhythm of suppression and, therefore, be more advantageous by restoring rhythmicity. Various circadian clock components include inherent small molecule ligands that frequently enhance their functions. Consequently, several drug-like compounds have been formulated to target a limited number of clock components and their regulators (refer to Table 8.1). Out of these substances, a small number have favourable bioavailability and pharmacodynamic qualities, making them suitable for pre-clinical in vivo animal research. Most in vivo research involving these compounds has focused on three major categories of disorders: cancer, inflammatory diseases, and metabolic diseases. It has been shown that administering these compounds to patients in all three disease groups has proven to be advantageous (Table 8.1). Furthermore, studies have shown that melanopsin antagonists are beneficial in reducing sensitivity to light (photophobia) (Jones et al. 2013). However, more research is necessary to determine the optimal use and enhance the pharmacological properties of this category of medications. Below we describe a few of the specific instances of drugging the clock components.

8.4.1 The Clock Gene: PAS/CLOCK: BMAL1

Due to their association with diseases, certain proteins have already been selected as targets for therapeutic screening. It has been demonstrated that this factor can bind tiny compounds in the PAS-B pocket, which function as inhibitors of the HIF2α: ARNT transcription factor (Baker et al. 2020). Although there have been successful discoveries of compounds that regulate other members of this family, few screenings have identified a promising candidate for CLOCK or BMAL1. One possible explanation for this is that the sizes of their PAS domain pockets are relatively tiny. The dimensions of the binding pocket partially determine its capacity to bind small

Table 8.1 Some small drug molecules which modulate the clock functioning

Drug/molecule name	Target clock component	Activity/mechanism
KS15	CRY inhibitor	Activation and enhancement of CLOCK-BMAL1-evoked E-box mediated transcription
Longdaysin	CKIα/δ inhibitor	The period-lengthening effect
DK359 (caged longdaysin)	Inactive against CKIα/δ	Temporal control of the clock period
LY2857785	CDK9 inhibitor	Enhancement of REV-ERBα expression
ARN5187	REV-ERBα/β agonist	Apoptotic induction
Nobiletin	RORs	Agonist of RORs
Digoxin	ROR-γ	Inverse agonist of *ROR-γ*
SR6542 (GSK4112)	REV-ERB agonist	Inhibition of Bmal1 expression
SR9009	REV-ERB agonist	Circadian clock repressor
MLN4924	RORα stabilization	Transactivation of Bmal1

molecules. A pocket with a diminutive size will have challenges in accommodating chemicals that are particular to proteins. The much smaller binding pockets of PAS-B could render CLOCK and BMAL1 more challenging targets compared to other bHLH-PAS proteins. Although the PAS pockets are small, both NPAS2 and CLOCK have been observed to bind the big molecule HEME. However, this binding is believed to be axial, which could explain why these pockets are able to hold a ligand of such size (Uchida et al. 2012). In addition, the AHR protein is stimulated by ligands, such as dioxins, that are smaller than HEME and the ligand binding domain of AHR is located within the PAS-B pocket (Fukunaga et al. 1995). Aside from the PAS-B pockets, there exist other locations within the CLOCK/BMAL1 structures that could potentially be influenced by small molecules. Several mutations have been documented that impede the process of CLOCK:BMAL1 dimerization, primarily located at the interface of the two proteins.

To effectively target the interface in the PAS domains, it appears that many mutations are necessary to achieve the desired outcome. This is likely because there is a significant interface between these domains. Although these mutations in the interface do indeed modify the interaction between CLOCK and BMAL1, it may be challenging to use them as targets for drug screening due to the proteins' flexibility and the requirement for them to be in monomeric state for effective targeting. Notwithstanding this difficulty, a recent study documents achievement in directing attention towards the junction between CLOCK and BMAL1 (Doruk et al. 2020). By employing an in silico screening technique, studies (Doruk et al. 2020) conducted a comprehensive search of the whole structure of CLOCK. They successfully docked two million molecules and discovered 100 compounds that exhibited predicted binding affinities within the range of −7 to −10 kcal/mol. CLK8 is anticipated to specifically attach to the bHLH region of CLOCK, rather than the PAS-B domain. When testing these chemicals in U2OS cells, researchers discovered that CLK8 interferes with the interaction between CLOCK and BMAL1, resulting in an increase in the circadian amplitude. However, it does not affect the length of the circadian period (Doruk et al. 2020).

8.4.2 The Clock Gene: Period

mPER1 and mPER2 are proteins consisting of 1290 and 1255 amino acids, respectively. They have two N-terminal PAS domains that are capable of facilitating interactions between proteins, such as homodimerization (interaction between two identical proteins) and heterodimerization (interaction between two different proteins) (Yagita et al. 2001). The protein contains a CRY binding domain (CBD) at its C-terminal end, which enables it to engage with CRY and form a repressive complex with CLOCK:BMAL1. The structure of the carboxy-terminal binding domain (CBD) of PER2, consisting of 128 amino acids, has been determined while it is bound to either CRY1 or CRY2 (Nangle et al. 2014). The CBD of PER2 interacts with CRY by encircling the protein near its secondary binding pocket. Additionally, there exists the Casein Kinase-Binding domain (CKBD), positioned between the

PAS domains and the CBD. This domain enables interactions with Casein kinaseδ/ε (CKδ/ε) (Lee et al. 2004), while its structure remains undisclosed. It has been well-established that mutations in the PAS domain of PER2 proteins can disrupt circadian rhythms in mice (Zheng et al. 2001). Recent research has demonstrated that even a single amino acid mutation in the PAS domain of PER2 can have an impact on circadian rhythmicity (Militi et al. 2016), highlighting the significance of the PAS domains in PER. It is tempting to imagine that the PAS domains in PER can be targeted similarly to the bHLH-PAS proteins to disrupt the protein-protein interaction. In contrast to the PAS domains, CBD is not a probable target due to its brevity and lack of patterns that could bind prospective chemicals. The CKBD may hold significant appeal, yet the domain's structure remains unclear, posing challenges in predicting drug binding.

8.4.3 The Clock Genes: Cryptochrome

Cryptochromes are flavoproteins that share structural similarities with photolyases, which are proteins involved in light-dependent DNA repair. However, it is important to note that cryptochrome lacks photolyase activity. There exist four distinct forms of CRYs, namely CRY1, CRY2, CRY4, and DASH-CRY. Contrary to the activators of the TTFL, CRYs have diverse functions. In Drosophila melanogaster, CRY functions as a light-sensitive regulator of circadian rhythms (Emery et al. 2000). In mammals, CRY1 and CRY2 are non-light-sensitive inhibitors of the TTFL (Griffin Jr et al. 1999). Recent evidence suggests that CRY4 has a crucial role as a magnetoreceptive protein in migratory birds, enabling them to detect magnetic fields (Xu et al. 2021).

CRY is composed of two domains: the Photolyase Homology Region (PHR) domain and the C-terminal tail. The PHR domain contains two distinct binding pockets: the primary and secondary. Conversely, the C-terminal domain lacks a defined structure. The main binding site is recognised for its ability to bind a FAD co-factor (for details, see (Cashmore 2003)) and also enable interactions with FBXL3/21. In photolyases, the secondary binding pocket is responsible for attaching a small molecule co-factor, such as methenyltetrahydrofolate, which functions as a light antennae. However, in mammalian CRYs, this pocket facilitates interactions with CLOCK through the CLOCK: TRP362 loop. While the PHR domains of CRY1 and CRY2 show a high degree of similarity, their disordered C-terminal tails are radically different, and the purpose of these tails is not well comprehended. The presence of two distinct binding spaces in CRYs renders them highly appealing as candidates for drug screening. The first CRY agonist (KL001) was discovered through a cell-based phenotypic screen specifically designed to identify drugs that modulate circadian rhythmicity. KL001 functions by suppressing the interaction of the ubiquitin ligases FBXL3 and FBXL21 with CRY (Yoo et al. 2013), which also attach to the main binding site of CRY and label it for destruction.

By binding tiny molecules in the binding pocket, the lifetime of CRY is extended, resulting in a prolonged repressive phase of the clock (Hirota and Kay 2015). By systematically investigating the structure activity relationship (SAR) starting from KL001, researchers were able to develop new compounds that exhibited greater potency and improved in vivo exposure. These compounds were found to be effective in vivo in enhancing glucose tolerance in the db/db mice model of diabetes (Humphries et al. 2016). Another research team employed in silico docking screens and molecular dynamics simulations to discover other chemicals that are anticipated to bind to the major pocket and induce an elongation of the period (Gul et al. 2021). The secondary pocket in CRY1 and CRY2 exhibits a high degree of homology and is distinguished by only 5 amino acids, all of which are conservative variations. Despite the significant resemblance, research has demonstrated that CRY1 can independently connect to and suppress CLOCK, whereas CRY2 needs PER2 to efficiently bind to CLOCK. Attaching small molecules to the secondary pocket can potentially prolong the transcriptional phase of the clock by inhibiting the suppression of CLOCK/BMAL1 activity, either at the start (CRY2) or end (CRY1) of the repressive phase. Due to the significant similarity in sequence between CRY1 and CRY2, it was challenging to create inhibitors that specifically target each isoform. However, researchers have successfully created various compounds, namely KL101 and KL201 for CRY1 and TH301 for CRY2, which exhibit selective binding and isoform-specific regulation (Miller et al. 2020). The crystal structures of CRY1 and CRY2, when bound to the chemicals, revealed an unexpected finding: they both interact with identical amino acid residues in the main binding site of each protein. This suggests that the specificity of the interaction is determined by the disordered and distinct C-terminal tails. Nevertheless, exchanging the C-terminal tails did not result in the reversal of specificity; instead, it led to the loss of binding. This suggests that a certain combination of the PHR and tail is crucial. Using enhanced crystal structures with greater electron density in the main pocket and molecular dynamics simulations, it was discovered that while the chemicals interact with identical residues in the pockets, the positioning of these residues differs between CRY1 and CRY2. This study discovered a crucial "gatekeeper" tryptophan in both proteins that interacts with a residue in the adjacent loop structure (known as the "lid loop") in different manners in the two proteins (Miller et al. 2020). This work suggests that the C-terminal tails may aid in facilitating these conformational changes. While the crystal structures of mammalian CRYs do not have disordered tails, in Drosophila, the CRY tail interacts with the same tryptophan and lid loop structure, providing evidence for this concept. Further phenotypic screenings have recently discovered specific agonists for CRY1, namely TH303 and TH129, which do not rely on the C-terminal tail for their specificity(Hirota and Kay 2015). The chemicals caused a rearrangement of the lid loop structure by interacting with F409 in the lid loop of CRY1 through the benzophenone moiety in the compounds.

The isoform-specific chemicals are highly valuable for their utility in analysing and differentiating the actions of the two CRY protein isoforms. By substituting the benzophenone moiety with a photo-switchable derivative called azobenzene, TH129 underwent further chemical alteration and gave rise to a new molecule known as GO1423 (Kolarski et al. 2021). Light enables the ability to change the length of circadian periods in a reversible manner and broadens the range of methods available for manipulating the function of CRY1 in both space and time. Modulators of CRY are being explored for their potential to target the circadian clock in glioblastoma tumours. These tumours exploit the circadian clock to stimulate growth in cancer stem cells. Additionally, these chemicals have significant value as tools and may have a role in metabolic control. It is anticipated that CRY stabilizers, which induce inhibition of CLOCK:BMAL1, will aid in the suppression of tumour growth. Furthermore, several research teams have discovered supplementary CRY modulators that function to mildly impede the interactions between CRY1 and CRY2 and BMAL1. Chun and colleagues (Chun et al. 2014) discovered KS15, a molecule that increased CLOCK/BMAL1 transcriptional activation by approximately two-fold, by conducting a screen of ~1000 drug-like compounds. Surprisingly, this inhibitor only slightly reduces the strength of the BMAL1-luciferase rhythms in cells and does not affect the length of the period (Jang et al. 2018).

8.4.4 The Clock Genes: Retinoic Acid Receptor-Related Orphan Receptors (Reverb and RoR)

REV-ERBα/β and RORα/β/γ belong to the nuclear receptor superfamily of transcription factors. They contribute to an auxiliary circadian loop that enhances the stability of the biological clock by regulating the expression of CLOCK and BMAL1 (Preitner et al. 2002; Sato et al. 2004). Both ROR and REV-ERBs exert their effects by acting on the same enhancer element. However, ROR stimulates transcription, while REV-ERBs inhibit transcription. REV-ERB antagonists are predicted to enhance circadian transcription, while ROR agonists are also likely to have the same effect. Similar to AHR, REV-ERBs and RORs are recognized to be regulated by ligands, with heme serving as an intrinsic ligand for REV-ERB (Raghuram et al. 2007). The specific molecules that naturally bind to RORs and regulate their activity are not fully comprehended, but sterols like cholesterol have been suggested as potential candidates. Attaching heme to REV-ERBs increases the ability to attract its corepressor complex, including nuclear corepressor (NCoR) and histone deacetylase 3 (HDAC3). After making this discovery, a biochemical test was conducted to find synthetic ligands. The test used a fluorescence resonance energy transfer (FRET) assay to measure the recruitment of NCoR. This led to the identification of the first synthetic agonist, called GSK4112 (Meng et al. 2008). Nevertheless, the compound's in vivo application was restricted due to its unfavourable pharmacokinetic (PK) features. Subsequent to that period, several substances that stimulate a response as well as a small number of substances that inhibit a response have been formulated and examined by multiple research teams (as discussed in (Uriz-Huarte et al. 2020)). While several of them have demonstrated encouraging outcomes in cellular and in vivo settings, there have been numerous challenges in attaining satisfactory bioavailability and other desirable pharmacokinetic profiles.

Additionally, the effects of certain substances differ across different cell lines and tissues. Furthermore, there have been reports of unintended effects of certain chemicals, known as off-target effects (Dierickx et al. 2022). Regrettably, there is a significant dearth of comprehensive research on the binding site or binding mechanism for a majority of these drugs, despite the utilisation of in silico docking and computational predictions (Westermaier et al. 2017). The initial discovery of a synthetic agonist for ROR was made with the diabetic medication thiazolidinedione (Missbach et al. 1996). Subsequently, an inverse agonist was also discovered for RORs (Kumar et al. 2012), which served as the foundation for the development of other agonists and inverse agonists. A separate chemical, named Nobiletin, was discovered by an impartial screening process for circadian effectors. This process involved a high throughput screening of 200,000 compounds utilising PER2-luciferase reporter cells. Nobiletin is a naturally occurring compound found in citrus peels. The deconvolution of its target demonstrated that it functions as an agonist of ROR and competes for binding to the ligand binding domains of both RORα and RORγ. It exhibits a greater affinity for RORγ. Nobiletin functions as a powerful regulator of circadian rhythm strength, leading to many advantageous impacts on metabolism and extended lifespan in mice ((Nohara et al. 2019) for further details).

8.4.5 The Kinases

Casein kinases are a group of enzymes that phosphorylate serine and threonine residues. They play a crucial role in controlling various biological processes, such as circadian rhythms, DNA repair, cell death, and cell specialisation. Additionally, they have been associated with the development of diseases such as cancer, Alzheimer's disease, and depression (Schittek and Sinnberg 2014). The Casein kinase family comprises CK1, which encompasses many isoforms, including CK1α/β/γ1/γ2/γ3/δ/ϵ, along with splice variants of these isoforms, and CK2. The involvement of CK proteins in circadian rhythms has been extensively studied, particularly CK1ε/δ and CK2. These proteins interact with PER1/2 and phosphorylate various locations on the protein, thereby regulating its stability (Philpott et al. 2023). In addition, CK1δ is a consistently present element in both the cytosolic CRY/PER and the nuclear CLOCK/BMAL1/CRY/PER megacomplexes.

The discovery of the period-lengthening molecule longdaysin (Hirota et al. 2010) was achieved using a high-throughput phenotypic screen, followed by chemical optimization of the identified hit compounds. Target deconvolution analysis demonstrated that longdaysin effectively bound to and reduced the activity of both CK1α and CK1δ, as well as ERK2. Expanding on the purine structure of longdaysin, a more powerful molecule called NCC007 was identified. A recent development involves the use of azobenzene derivatives of longdaysin to create a photosensitive switchable CK1 inhibitor. This inhibitor enables the manipulation of the period duration by reversible inhibition of the kinase, allowing for precise control over both the spatial and temporal aspects of the process (Kolarski et al. 2021).

Furthermore, because CK1 proteins have a wide-ranging impact on several diseases, numerous CK1 inhibitors have been created for purposes unrelated to the circadian clock domain. These inhibitors exhibit different levels of specificity and unintended effects on other targets. Circadian biologists have utilised these chemicals to investigate the functions of the CK1 enzymes in the circadian clock. Pfizer initially discovered PF670462, a potent inhibitor of CKIε/δ, which was found to induce phase-delaying effects in mice (Badura et al. 2007). Additional investigations led to the discovery of a molecule called PF4800567, which selectively targets a specific isoform, CKIε, with a 22-fold higher affinity (Walton et al. 2009). These two chemicals were used to demonstrate that inhibiting CK1ε alone had negligible impact on the duration of the circadian period. However, inhibiting CK1δ resulted in significantly longer period lengths in both cultured cells and live animals (Walton et al. 2009). By using inhibitors and specific genetic knockouts, it was determined that the CK1δ isoform is primarily responsible for determining the period. Utilising isoform-specific inhibitors exemplifies how targeted modulators can serve as effective tools for analysing the intricate molecular mechanics of the clock. Recently, a screening method using cells has successfully identified two CK2 inhibitors, CX-4945 and GO289. These inhibitors have been found to increase the duration of the circadian rhythm in U2OS cells and also hinder the growth of human renal cell carcinoma and mouse leukaemia cells (Oshima et al. 2019). Undoubtedly, these novel methods will uncover fresh insights into the role of CK2 in the circadian clock.

8.5 Effects of Physical Activity on Circadian Rhythms in the Elderly

Human aging is characterized by a gradual and widespread deterioration. To mitigate the negative effects of reduced capacity for self-sufficiency in day-to-day activities, it is recommended to engage in regular physical exercise (WHO 2010). The process of aging is linked to a disturbance in the chronobiological cycle (Touitou and Haus 2000), resulting in a decline in the integrity of biological rhythmicity, including temperature regulation, sleep-wake patterns, and alertness (Dijk et al. 2000; Hofman 2000). Exploring methods to enhance the circadian timing system could be beneficial for the overall health and wellness of older individuals. Specifically, the changes in biological rhythms that occur with aging can be attributed, in part, to the sedentary habits of older individuals (Van Someren 2000). The emerging idea is that physical activity serves as a non-photic zeitgeber, or external cue, for the internal clock.

8.5.1 Physical Activity as a Modulator of Circadian Rhythm

Several agents that affect the circadian timing system, such as melatonin (Zhdanova and Wurtman 1997), intense light (Turner et al. 2010), and alterations in body temperature (Van Someren 2000), have been successfully studied to restore circadian rhythmicity. A recent concept suggests that physical activity may serve as a powerful regulator of the biological clock.

There is currently enough data from phase-shift studies to create a phase-response curve that compares the onset of dim light melatonin to acute exercise in people.

This phase-response curve incorporates data from studies that utilised either 1-h high-intensity or 3-h lower-intensity workouts on exercise machines. The findings indicate that exercise in the evening has a phase-advancing effect, occurring before the usual onset of sleep and the onset of dim light melatonin. On the other hand, exercise during most of the typical sleep period, extending into the morning and potentially the afternoon, has a phase-delaying effect (for details, see Baehr et al. 1999). The studies utilised in this analysis were limited by a lack of control for competing zeitgebers, such as light.

Nevertheless, the phase-response curve's general form is distinct enough from the photic curve to eliminate the possibility that light exposure during exercise is a probable cause of phase shifts. This is especially true for advances in response to evening exercise and delays in response to late morning exercise, which are phases where light would induce shifts in the opposite direction (Mistlberger and Skene 2005). The impact of exercise sessions was observable during the initial circadian cycle following the session but not during the subsequent cycle (Buxton et al. 2003), indicating that there was no consistent adjustment of the pacemaker phase.

Due to the ability of physical activity to cause a change in the start of dim light melatonin, several researchers have examined the positive impact of exercise in speeding up the process of resynchronizing the body after experiencing jetlag (Barger et al. 2004). The study found that exercising in dim light circumstances caused a significant shift in the commencement of melatonin production by 3.17 h. Nevertheless, it is possible to create a prolonged alteration in phase shift by exposing oneself to light at the right time (Duffy et al. 2001).

Investigations (Baehr et al. 1999; Youngstedt et al. 2002), have investigated the impact of acute exercise on the circadian rhythm of core temperature in humans. This research specifically explored how exercise interacts with bright light and circadian rhythms. The concurrent application of intense illumination and physical activity resulted in a notable shift in the timing of a biological rhythm. However, this shift was not statistically distinguishable from the one induced alone by intense illumination. The researchers reached the conclusion that their study lacked sufficient statistical power (Youngstedt et al. 2002) or did not have enough intensity and duration of exercise. Current research on acute exercise has predominantly focused on young adult males.

Interestingly, the long-term impacts of exercise did not provide evidence to support the occurrence of phase shifts (Atkinson et al. 2007). There was no difference between the groups in terms of the synchronization of the rhythms. Nevertheless, no research has been conducted to explore the correlation between the timing of training and the circadian rhythms of those who engage in physical activity.

Hence, the impact of physical activity on the alteration of circadian rhythms' phase remains ambiguous in humans. Applied research has been negatively impacted by the absence of control over competing zeitgebers, such as light, as well as the exercise characteristics of research participants, including mode, intensity, duration, and athletic status. Standardization of these variables is necessary to offer evidence-based recommendations for including physical activity in resynchronization methods.

8.5.2 The Effect of Physical Status on Circadian Rhythm of Elderly

It is well-established that lack of physical exercise can result in many illnesses related to the rhythm of rest and activity, as well as a reduction in the amplitude of body temperature and the regular secretion of hormones (Hackney and Lane 2015). The decline in daily physical activity among older adults is likely to contribute to changes in circadian rhythmicity. The correlation between physical exercise and physical capacity is widely recognized. Regarding the connection between physical activity and physical ability, the changes in circadian rhythmicity in older persons may also rely on their aerobic capacity (Milanović et al. 2013). A study examining the correlation between aerobic ability and circadian rhythm in elderly individuals in which each participant underwent an individual assessment of circadian rhythmicity traits, which were determined by analysing their core temperature circadian rhythms (Dupont Rocher et al. 2016). Actigraphy was used to measure sleep during the night and activity during the day. Additionally, a simulated driving task was used to assess alertness during the day. Typically, as individuals age, a reduction in the intensity of many circadian rhythms, such as the core temperature rhythm, is commonly observed (Hanneman 2001). Nevertheless, there are other arguments that support the rejection of this idea.

Older participants with a higher level of physical capacity appear to have an augmented biological clock. Physical exercise has a direct or indirect influence on the circadian clock, which is the primary mechanism of biological rhythm. As a result, it counteracts the negative effects of circadian clock aging. Nevertheless, aerobic capacity may also serve as an indicator of overall health condition, and the youthful nature of the biological rhythm found in groups with high aerobic capacity may rely on the participants' overall health status. Researchers have done experiments incorporating a physical training routine to partially mitigate this bias. Prior research on the impact of enhanced physical fitness on circadian rhythms in elderly individuals is scarce. The study (Van Someren 2000) examined whether long-term fitness training may enhance the circadian rest-activity rhythm in healthy older individuals. To achieve this objective, a group of elderly males in good health took part in a training programme lasting three months. The implementation of fitness training resulted in a notable decrease in the disruption of the rest-activity rhythm following the training period. It is noteworthy that none of the participants in this study reported making any lifestyle modifications in terms of their overall involvement in activities or time spent outside. Hence, the outcome cannot be accounted for by the influence of other external synchronizers, such as light. In contrast, studies have found a notable impact of 4 months of physical training on both the magnitude of the activity-rest rhythm and the core temperature rhythms. These observations are corroborated by the evidence that exercise training in older individuals can improve sleep quality and daytime alertness (Gruau et al. 2003; Montgomery and Dennis 2004). The research mentioned above only included a limited number of studies. This may account for the absence of uniformity in the reported findings. While additional research is required, the theory suggesting that physical exercise has a beneficial impact on circadian rhythmicity appears valid presently.

8.6 Conclusion

In recent years, there have been significant studies in the field of chronobiology that have enhanced our comprehension of the functions and mechanisms of the circadian clock in relation to human health and disorders, such as cancer. The disturbance of circadian rhythms has a detrimental effect on both the internal clocks of tumours and the biological systems of the host, leading to an increased risk and progression of cancer. However, due to the strong association between cancers and other conditions that disrupt the body's natural rhythms, such as aging, obesity, and diabetes, it is anticipated that improving circadian rhythms through interventions that are based on the body's internal clock (such as chronophototherapy, chronodiet, and chrono-exercise) will greatly improve overall circadian health. Moreover, there are still obstacles in effectively incorporating the knowledge of circadian rhythms into other areas of healthcare. It is necessary to persuade pharmaceutical corporations and healthcare practitioners that circadian rhythm is not a hindrance but rather a chance to deliver patient-centred care. The advancement in the field of post-genomics during the past 20 years has resulted in gradual progress in the treatment of several chronic diseases, including cancer. Regarding other medical conditions like arthritis, it is believed that we have reached a point where we can provide care at a fair cost, although the disease has not been adequately addressed. Delayed-release or slow-release formulations and distribution using smart pumps provide opportunities for innovation and market expansion in the treatment of such disorders. Evaluating the pharmacodynamic and pharmacokinetic (PD/PK) properties of investigational new medications in live organisms, considering factors such as circadian rhythm, timing of meals, sleep, light, and activity, can greatly enhance the accuracy and effectiveness of phase I and subsequent clinical trials, hence improving the drug development process.

The integration of conventional nutritional guidelines with circadian lifestyles, which involve the synchronisation of sleep, physical activity, dietary habits, and illumination, has the potential to enhance prehabilitation, prognosis, and rehabilitation outcomes for patients. Particularly, those receiving intense medical treatments like chemotherapy, radiation therapy, or operations may find advantages in adopting a lifestyle that is specifically adjusted to their circadian rhythm. In summary, the enthusiasm surrounding circadian rhythm and its importance in preserving the well-being of various organ systems can be encapsulated in the concept of the "Circadian theory of health". The "Circadian theory of health" has the potential to have a significant effect on healthcare by reducing fatalities from infectious diseases and improving the longevity of persons.

References

Acosta-Rodríguez VA, Rijo-Ferreira F, Green CB, Takahashi JS (2021) Importance of circadian timing for aging and longevity. Nat Commun 12(1):2862

Allada R, Bass J (2021) Circadian mechanisms in medicine. N Engl J Med 384(6):550–561

Antoch MP, Kondratov RV (2013) Pharmacological modulators of the circadian clock as potential therapeutic drugs: focus on genotoxic/anticancer therapy. Circadian Clocks (217):289–309

Atkinson G, Edwards B, Reilly T, Waterhouse J (2007) Exercise as a synchroniser of human circadian rhythms: an update and discussion of the methodological problems. Eur J Appl Physiol 99:331–341

Badura L, Swanson T, Adamowicz W, Adams J, Cianfrogna J, Fisher K, Holland J, Kleiman R, Nelson F, Reynolds L (2007) An inhibitor of casein kinase Iε induces phase delays in circadian rhythms under free-running and entrained conditions. J Pharmacol Exp Ther 322(2):730–738

Baehr EK, Fogg LF, Eastman CI (1999) Intermittent bright light and exercise to entrain human circadian rhythms to night work. Am J Phys Regul Integr Comp Phys 277(6):R1598–R1604

Baker JR, Sakoff JA, McCluskey A (2020) The aryl hydrocarbon receptor (AhR) as a breast cancer drug target. Med Res Rev 40(3):972–1001. https://doi.org/10.1002/med.21645

Banerjee S, Wang Y, Solt LA, Griffett K, Kazantzis M, Amador A, El-Gendy BM, Huitron-Resendiz S, Roberts AJ, Shin Y (2014) Pharmacological targeting of the mammalian clock regulates sleep architecture and emotional behaviour. Nat Commun 5(1):5759

Banks G, Nolan PM, Peirson SN (2016) Reciprocal interactions between circadian clocks and aging. Mammalian Genome 27(7–8):332–340. https://doi.org/10.1007/s00335-016-9639-6

Barger LK, Wright KP Jr, Hughes RJ, Czeisler CA (2004) Daily exercise facilitates phase delays of circadian melatonin rhythm in very dim light. Am J Phys Regul Integr Comp Phys 286(6):R1077–R1084

Bersten DC, Sullivan AE, Peet DJ, Whitelaw ML (2013) bHLH–PAS proteins in cancer. Nat Rev Cancer 13(12):827–841

Bryant Z, Oberstrass FC, Basu A (2012) Recent developments in single-molecule DNA mechanics. Curr Opin Struct Biol 22(3):304–312

Buxton OM, Lee CW, L'Hermite-Balériaux M, Turek FW, Van Cauter E (2003) Exercise elicits phase shifts and acute alterations of melatonin that vary with circadian phase. Am J Phys Regul Integr Comp Phys 284(3):R714–R724

Cashmore AR (2003) Cryptochromes: enabling plants and animals to determine circadian time. Cell 114(5):537–543

Chun SK, Jang J, Chung S, Yun H, Kim N-J, Jung J-W, Son GH, Suh Y-G, Kim K (2014) Identification and validation of cryptochrome inhibitors that modulate the molecular circadian clock. ACS Chem Biol 9(3):703–710. https://doi.org/10.1021/cb400752k

Cohen AA, Milot E, Yong J, Seplaki CL, Fülöp T, Bandeen-Roche K, Fried LP (2013) A novel statistical approach shows evidence for multi-system physiological dysregulation during aging. Mech Ageing Dev 134(3–4):110–117

Dierickx P, Zhu K, Carpenter BJ, Jiang C, Vermunt MW, Xiao Y, Luongo TS, Yamamoto T, Martí-Pàmies Í, Mia S (2022) Circadian REV-ERBs repress E4bp4 to activate NAMPT-dependent NAD+ biosynthesis and sustain cardiac function. Nat Cardiovasc Res 1(1):45–58

Dijk D-J, Duffy JF, Czeisler CA (2000) Contribution of circadian physiology and sleep homeostasis to age-related changes in human sleep. Chronobiol Int 17(3):285–311

Doruk YU, Yarparvar D, Akyel YK, Gul S, Taskin AC, Yilmaz F, Baris I, Ozturk N, Türkay M, Ozturk N (2020) A CLOCK-binding small molecule disrupts the interaction between CLOCK and BMAL1 and enhances circadian rhythm amplitude. J Biol Chem 295(11):3518–3531

Duffy JF, Rimmer DW, Czeisler CA (2001) Association of intrinsic circadian period with morningness–eveningness, usual wake time, and circadian phase. Behav Neurosci 115(4):895

Dupont Rocher S, Bessot N, Sesboüé B, Bulla J, Davenne D (2016) Circadian characteristics of older adults and aerobic capacity. J Gerontol Ser A 71(6):817–822

Emery P, Stanewsky R, Helfrich-Förster C, Emery-Le M, Hall JC, Rosbash M (2000) Drosophila CRY is a deep brain circadian photoreceptor. Neuron 26(2):493–504

Fribourgh JL, Srivastava A, Sandate CR, Michael AK, Hsu PL, Rakers C, Nguyen LT, Torgrimson MR, Parico GCG, Tripathi S (2020) Dynamics at the serine loop underlie differential affinity of cryptochromes for CLOCK: BMAL1 to control circadian timing. elife 9:e55275

Fukunaga BN, Probst MR, Reisz-Porszasz S, Hankinson O (1995) Identification of functional domains of the aryl hydrocarbon receptor. J Biol Chem 270(49):29270–29278

Gallego M, Virshup DM (2007) Post-translational modifications regulate the ticking of the circadian clock. Nat Rev Mol Cell Biol 8(2):139–148

Gill S, Le HD, Melkani GC, Panda S (2015) Time-restricted feeding attenuates age-related cardiac decline in drosophila. Science 347(6227):1265–1269

Gorbacheva VY, Kondratov RV, Zhang R, Cherukuri S, Gudkov AV, Takahashi JS, Antoch MP (2005) Circadian sensitivity to the chemotherapeutic agent cyclophosphamide depends on the functional status of the CLOCK/BMAL1 transactivation complex. Proc Natl Acad Sci 102(9):3407–3412

Griffin EA Jr, Staknis D, Weitz CJ (1999) Light-independent role of CRY1 and CRY2 in the mammalian circadian clock. Science 286(5440):768–771

Gruau S, Pottier A, Davenne D, Denise P (2003) Les facteurs d'accidents de la route par somnolence chez les conducteurs âgésPrévention par l'activité physiqueSleepiness-related road traffic accidents among elderly drivers: an analysis of the contributory factorsThe preventive role of physical activity. Recherche Transport Sécurité 79–80:134–144. https://doi.org/10.1016/S0761-8980(03)00011-6

Gul S, Rahim F, Isin S, Yilmaz F, Ozturk N, Turkay M, Kavakli IH (2021) Structure-based design and classifications of small molecules regulating the circadian rhythm period. Sci Rep 11(1):18510

Hackney AC, Lane AR (2015) Exercise and the regulation of endocrine hormones. Prog Mol Biol Transl Sci 135:293–311

Hanneman SK (2001) Measuring circadian temperature rhythm. Biol Res Nurs 2(4):236–248

Haus E, Halberg F, Scheving LE, Pauly JE, Cardoso S, Kühl JF, Sothern RB, Shiotsuka RN, Hwang DS (1972) Increased tolerance of leukemic mice to arabinosyl cytosine with schedule adjusted to circadian system. Science 177(4043):80–82

He B, Chen Z (2016) Molecular targets for small-molecule modulators of circadian clocks. Curr Drug Metab 17(5):503–512

Helleboid S, Haug C, Lamottke K, Zhou Y, Wei J, Daix S, Cambula L, Rigou G, Hum DW, Walczak R (2014) The identification of naturally occurring neoruscogenin as a bioavailable, potent, and high-affinity agonist of the nuclear receptor RORα (NR1F1). J Biomol Screen 19(3):399–406

Hirota T, Kay SA (2009) High-throughput screening and chemical biology: new approaches for understanding circadian clock mechanisms. Chem Biol 16(9):921–927

Hirota T, Kay SA (2015) Identification of small-molecule modulators of the circadian clock. In: Methods in enzymology, vol 551. Elsevier, pp 267–282

Hirota T, Lee JW, Lewis WG, Zhang EE, Breton G, Liu X, Garcia M, Peters EC, Etchegaray J-P, Traver D (2010) High-throughput chemical screen identifies a novel potent modulator of cellular circadian rhythms and reveals CKIα as a clock regulatory kinase. PLoS Biol 8(12):e1000559

Hofman MA (2000) The human circadian clock and aging. Chronobiol Int 17(3):245–259

Huh JR, Leung MW, Huang P, Ryan DA, Krout MR, Malapaka RR, Chow J, Manel N, Ciofani M, Kim SV (2011) Digoxin and its derivatives suppress TH17 cell differentiation by antagonizing RORγt activity. Nature 472(7344):486–490

Humphries PS, Bersot R, Kincaid J, Mabery E, McCluskie K, Park T, Renner T, Riegler E, Steinfeld T, Turtle ED (2016) Carbazole-containing sulfonamides and sulfamides: discovery of cryptochrome modulators as antidiabetic agents. Bioorg Med Chem Lett 26(3):757–760

Isojima Y, Nakajima M, Ukai H, Fujishima H, Yamada RG, Masumoto K, Kiuchi R, Ishida M, Ukai-Tadenuma M, Minami Y (2009) CKIε/δ-dependent phosphorylation is a temperature-insensitive, period-determining process in the mammalian circadian clock. Proc Natl Acad Sci 106(37):15744–15749

Jagannath A, Taylor L, Wakaf Z, Vasudevan SR, Foster RG (2017) The genetics of circadian rhythms, sleep and health. Hum Mol Genet 26(R2):R128–R138

Jang J, Chung S, Choi Y, Lim HY, Son Y, Chun SK, Son GH, Kim K, Suh Y-G, Jung J-W (2018) The cryptochrome inhibitor KS15 enhances E-box-mediated transcription by disrupting the feedback action of a circadian transcription-repressor complex. Life Sci 200:49–55

Jetten AM, Kang HS, Takeda Y (2013) Retinoic acid-related orphan receptors α and γ: key regulators of lipid/glucose metabolism, inflammation, and insulin sensitivity. Front Endocrinol 4:1

Jones KA, Hatori M, Mure LS, Bramley JR, Artymyshyn R, Hong S-P, Marzabadi M, Zhong H, Sprouse J, Zhu Q (2013) Small-molecule antagonists of melanopsin-mediated phototransduction. Nat Chem Biol 9(10):630–635

Katewa SD, Akagi K, Bose N, Rakshit K, Camarella T, Zheng X, Hall D, Davis S, Nelson CS, Brem RB (2016) Peripheral circadian clocks mediate dietary restriction-dependent changes in lifespan and fat metabolism in drosophila. Cell Metab 23(1):143–154

Koike N, Yoo S-H, Huang H-C, Kumar V, Lee C, Kim T-K, Takahashi JS (2012) Transcriptional architecture and chromatin landscape of the Core circadian clock in mammals. Science 338(6105):349–354. https://doi.org/10.1126/science.1226339

Kojetin DJ, Burris TP (2014) REV-ERB and ROR nuclear receptors as drug targets. Nat Rev Drug Discov 13(3):197–216

Kolarski D, Miró-Vinyals C, Sugiyama A, Srivastava A, Ono D, Nagai Y, Iida M, Itami K, Tama F, Szymanski W (2021) Reversible modulation of circadian time with chronophotopharmacology. Nat Commun 12(1):3164

Kumar N, Lyda B, Chang MR, Lauer JL, Solt LA, Burris TP, Kamenecka TM, Griffin PR (2012) Identification of SR2211: A potent synthetic RORγ-selective modulator. ACS Chem Biol 7(4):672–677

Lee C, Weaver DR, Reppert SM (2004) Direct association between mouse PERIOD and CKIε is critical for a functioning circadian clock. Mol Cell Biol 24:584

Levi F, Schibler U (2007) Circadian rhythms: mechanisms and therapeutic implications. Annu Rev Pharmacol Toxicol 47:593–628

Liu AC, Welsh DK, Ko CH, Tran HG, Zhang EE, Priest AA, Buhr ED, Singer O, Meeker K, Verma IM (2007) Intercellular coupling confers robustness against mutations in the SCN circadian clock network. Cell 129(3):605–616

Longo VD, Panda S (2016) Fasting, circadian rhythms, and time-restricted feeding in healthy lifespan. Cell Metab 23(6):1048–1059

Martin GM (2011) The biology of aging: 1985–2010 and beyond. FASEB J 25(11):3756

Meng QJ, McMaster A, Beesley S, Lu WQ, Gibbs J, Parks D, Collins J, Farrow S, Donn R, Ray D (2008) Ligand modulation of REV-ERBα function resets the peripheral circadian clock in a phasic manner. J Cell Sci 121(21):3629–3635

Milanović Z, Pantelić S, Trajković N, Sporiš G, Kostić R, James N (2013) Age-related decrease in physical activity and functional fitness among elderly men and women. Clin Interv Aging 8:549–556. https://doi.org/10.2147/CIA.S44112

Militi S, Maywood ES, Sandate CR, Chesham JE, Barnard AR, Parsons MJ, Vibert JL, Joynson GM, Partch CL, Hastings MH (2016) Early doors (Edo) mutant mouse reveals the importance of period 2 (PER2) PAS domain structure for circadian pacemaking. Proc Natl Acad Sci 113(10):2756–2761

Miller S, Aikawa Y, Sugiyama A, Nagai Y, Hara A, Oshima T, Amaike K, Kay SA, Itami K, Hirota T (2020) An isoform-selective modulator of cryptochrome 1 regulates circadian rhythms in mammals. Cell Chem Biol 27(9):1192–1198

Missbach M, Jagher B, Sigg I, Nayeri S, Carlberg C, Wiesenberg I (1996) Thiazolidine diones, specific ligands of the nuclear receptor retinoid Z receptor/retinoid acid receptor-related orphan receptor α with potent antiarthritic activity. J Biol Chem 271(23):13515–13522

Mistlberger RE, Skene DJ (2005) Non-photic entrainment in humans? J Biol Rhythm 20(4):339–352

Montgomery P, Dennis J (2004) A systematic review of non-pharmacological therapies for sleep problems in later life. Sleep Med Rev 8(1):47–62

Nangle SN, Rosensweig C, Koike N, Tei H, Takahashi JS, Green CB, Zheng N (2014) Molecular assembly of the period-cryptochrome circadian transcriptional repressor complex. elife 3:e03674
Nohara K, Yoo S-H, Chen Z (2015) Manipulating the circadian and sleep cycles to protect against metabolic disease. Front Endocrinol 6:35
Nohara K, Mallampalli V, Nemkov T, Wirianto M, Yang J, Ye Y, Sun Y, Han L, Esser KA, Mileykovskaya E (2019) Nobiletin fortifies mitochondrial respiration in skeletal muscle to promote healthy aging against metabolic challenge. Nat Commun 10(1):3923
Oshima T, Niwa Y, Kuwata K, Srivastava A, Hyoda T, Tsuchiya Y, Kumagai M, Tsuyuguchi M, Tamaru T, Sugiyama A (2019) Cell-based screen identifies a new potent and highly selective CK2 inhibitor for modulation of circadian rhythms and cancer cell growth. Sci Advances 5(1):eaau9060
Philpott JM, Freeberg AM, Park J, Lee K, Ricci CG, Hunt SR, Narasimamurthy R, Segal DH, Robles R, Cai Y (2023) PERIOD phosphorylation leads to feedback inhibition of CK1 activity to control circadian period. Mol Cell 83(10):1677–1692
Pierpaoli W, Regelson W (1994) Pineal control of aging: effect of melatonin and pineal grafting on aging mice. Proc Natl Acad Sci 91(2):787–791
Poon LW, Perls TT (2007) Annual review of gerontology and geriatrics: biopsychosocial approaches to longevity, vol 27. Springer, Cham
Preitner N, Damiola F, Zakany J, Duboule D, Albrecht U, Schibler U (2002) The orphan nuclear receptor REV-ERBα controls circadian transcription within the positive limb of the mammalian circadian oscillator. Cell 110(2):251–260
Raghuram S, Stayrook KR, Huang P, Rogers PM, Nosie AK, McClure DB, Burris LL, Khorasanizadeh S, Burris TP, Rastinejad F (2007) Identification of heme as the ligand for the orphan nuclear receptors REV-ERBα and REV-ERBβ. Nat Struct Mol Biol 14(12):1207–1213
Riemersma-van Der Lek RF, Swaab DF, Twisk J, Hol EM, Hoogendijk WJ, Van Someren EJ (2008) Effect of bright light and melatonin on cognitive and noncognitive function in elderly residents of group care facilities: a randomized controlled trial. JAMA 299(22):2642–2655
Rose MR (1994) Evolutionary biology of aging. Oxford University Press, Oxford
Rosensweig C, Green CB (2020) Periodicity, repression, and the molecular architecture of the mammalian circadian clock. Eur J Neurosci 51(1):139–165
Sancar A, Van Gelder RN (2021) Clocks, cancer, and chronochemotherapy. Science 371(6524):eabb0738
Santori FR (2015) Nuclear hormone receptors put immunity on sterols. Eur J Immunol 45(10):2730–2741
Sato TK, Panda S, Miraglia LJ, Reyes TM, Rudic RD, McNamara P, Naik KA, FitzGerald GA, Kay SA, Hogenesch JB (2004) A functional genomics strategy reveals Rora as a component of the mammalian circadian clock. Neuron 43(4):527–537
Schittek B, Sinnberg T (2014) Biological functions of casein kinase 1 isoforms and putative roles in tumorigenesis. Mol Cancer 13:1–14
Schroder EA, Esser KA (2013) Circadian rhythms, skeletal muscle molecular clocks, and exercise. Exerc Sport Sci Rev 41(4):224–229
Solt LA, Wang Y, Banerjee S, Hughes T, Kojetin DJ, Lundasen T, Shin Y, Liu J, Cameron MD, Noel R (2012) Regulation of circadian behaviour and metabolism by synthetic REV-ERB agonists. Nature 485(7396):62–68
Touitou Y, Haus E (2000) Alterations with aging of the endocrine and neuroendocrine circadian system in humans. Chronobiol Int 17(3):369–390
Turner PL, Van Someren EJ, Mainster MA (2010) The role of environmental light in sleep and health: effects of ocular aging and cataract surgery. Sleep Med Rev 14(4):269–280
Uchida T, Sagami I, Shimizu T, Ishimori K, Kitagawa T (2012) Effects of the bHLH domain on axial coordination of heme in the PAS-A domain of neuronal PAS domain protein 2 (NPAS2): conversion from His119/Cys170 coordination to His119/His171 coordination. J Inorg Biochem 108:188–195

Uriz-Huarte A, Date A, Ang H, Ali S, Brady HJ, Fuchter MJ (2020) The transcriptional repressor REV-ERB as a novel target for disease. Bioorg Med Chem Lett 30(17):127395

Van Someren EJ (2000) Circadian rhythms and sleep in human aging. Chronobiol Int 17(3):233–243

Vitaterna MH, Selby CP, Todo T, Niwa H, Thompson C, Fruechte EM, Hitomi K, Thresher RJ, Ishikawa T, Miyazaki J (1999) Differential regulation of mammalian period genes and circadian rhythmicity by cryptochromes 1 and 2. Proc Natl Acad Sci 96(21):12114–12119

Wallach T, Kramer A (2015) Chemical chronobiology: toward drugs manipulating time. FEBS Lett 589(14):1530–1538

Walton KM, Fisher K, Rubitski D, Marconi M, Meng Q-J, Sladek M, Adams J, Bass M, Chandrasekaran R, Butler T (2009) Selective inhibition of casein kinase 1ϵ minimally alters circadian clock period. J Pharmacol Exp Ther 330(2):430–439

Westermaier Y, Ruiz-Carmona S, Theret I, Perron-Sierra F, Poissonnet G, Dacquet C, Boutin JA, Ducrot P, Barril X (2017) Binding mode prediction and MD/MMPBSA-based free energy ranking for agonists of REV-ERBα/NCoR. J Comput Aided Mol Des 31:755–775

Witkowski S, Hagberg JM (2007) Progenitor cells and age: can we fight aging with exercise? J Appl Physiol 102(3):834–835

Wu D, Rastinejad F (2017) Structural characterization of mammalian bHLH-PAS transcription factors. Curr Opin Struct Biol 43:1–9

Xing W, Busino L, Hinds TR, Marionni ST, Saifee NH, Bush MF, Pagano M, Zheng N (2013) SCFFBXL3 ubiquitin ligase targets cryptochromes at their co-factor pocket. Nature 496(7443):64–68

Xu J, Jarocha LE, Zollitsch T, Konowalczyk M, Henbest KB, Richert S, Golesworthy MJ, Schmidt J, Déjean V, Sowood DJ (2021) Magnetic sensitivity of cryptochrome 4 from a migratory songbird. Nature 594(7864):535–540

Yagita K, Tamanini F, Van Der Horst GT, Okamura H (2001) Molecular mechanisms of the biological clock in cultured fibroblasts. Science 292(5515):278–281

Yoo S-H, Mohawk JA, Siepka SM, Shan Y, Huh SK, Hong H-K, Kornblum I, Kumar V, Koike N, Xu M (2013) Competing E3 ubiquitin ligases govern circadian periodicity by degradation of CRY in nucleus and cytoplasm. Cell 152(5):1091–1105

Youngstedt SD, Kripke DF, Elliott JA (2002) Circadian phase-delaying effects of bright light alone and combined with exercise in humans. Am J Phys Regul Integr Comp Phys 282(1):R259–R266

Zhang R, Lahens NF, Ballance HI, Hughes ME, Hogenesch JB (2014) A circadian gene expression atlas in mammals: implications for biology and medicine. Proc Natl Acad Sci 111(45):16219–16224

Zhao X, Hirota T, Han X, Cho H, Chong L-W, Lamia K, Liu S, Atkins AR, Banayo E, Liddle C (2016) Circadian amplitude regulation via FBXW7-targeted REV-ERBα degradation. Cell 165(7):1644–1657

Zhdanova IV, Wurtman RJ (1997) The pineal hormone-melatonin. In: Endocrinology: basic and clinical principles. Springer, Cham, pp 279–290

Zheng B, Albrecht U, Kaasik K, Sage M, Lu W, Vaishnav S, Li Q, Sun ZS, Eichele G, Bradley A (2001) Nonredundant roles of the mPer1 and mPer2 genes in the mammalian circadian clock. Cell 105(5):683–694

Zhu L, Zou F, Yang Y, Xu P, Saito K, Othrell Hinton A Jr, Yan X, Ding H, Wu Q, Fukuda M (2015) Estrogens prevent metabolic dysfunctions induced by circadian disruptions in female mice. Endocrinology 156(6):2114–2123

SPRINGER NATURE

GPSR Compliance

The European Union's (EU) General Product Safety Regulation (GPSR) is a set of rules that requires consumer products to be safe and our obligations to ensure this.

If you have any concerns about our products, you can contact us on ProductSafety@springernature.com

In case Publisher is established outside the EU, the EU authorized representative is:

Springer Nature Customer Service Center GmbH
Europaplatz 3
69115 Heidelberg, Germany

The manufacturer's authorised representative in the EU is Springer Nature Customer Service Centre GmbH, Europaplatz 3, 69115 Heidelberg, Germany. If you have any concerns regarding our products, please contact ProductSafety@springernature.com

Printed and bound by CPI Group (UK) Ltd, Croydon, CR0 4YY
01/10/2026
02225857-0001